Earned Hours© Presents

Weekly

Performance Cash

as **Controlled** by
the
Break Even for Each Hour Paid
that takes the mystery out of
Profit & Loss

By

Martin D'Amico

Author of How to Predict Year End Cash
From 30 years in the trenches of
Public & Cost Accounting
As a CPA & Corporate Controller

Hands On Programs For
CEOs & CFOs
Investment Bankers, Hospitality Managers
Retailers, Manufacturers
Public & Cost Accountants
&
An Open Letter to Shareholders

ISBN: 1-4033-6861-9 (e-book)
ISBN: 1-4033-6862-7 (Paperback)

Library of Congress Control Number: 2002111837

This book is printed on acid free paper.

Printed in the United States of America
Bloomington, IN

1stBooks - rev. 1/29/03

Preface

Inspired by the Sad State of Financial Reporting

Martin D'Amico sees a Business

Exposed by Four Benchmarks that gives a Business life,

Accounting only sees a Business in Conformity to Standards.

Quarterly dollars & cents of EPS have no accounting for cause & effect in pennies.

Weekly Performance Cash, an hourly rate in dollars & cents

Has cause & effect in pennies.

The Perfect Business Control

This book is intended to be a self contained document for completely understanding Earned Hours© & its Utilization (not a theory) without the "hands on" Diskette Programs that support the entire text. Accordingly & independent of the Diskette with only basic spreadsheet skills, the **XL 3 graphic** sheet from the Appendix has been presented for your immediate understanding & use. Much like assembling the purchase of a product before reading the instructions. If your skills do not prove out after experimenting with my data on your worksheet, I will be glad to personally correct the sheet when submitted to me in Excel via www.performancecash.com.

That one worksheet is also **"the book in a nutshell" & the perfect control for any business.** The sheet tells the irrefutable **Truth** for each hour paid to all your employees as to why you exceeded or did not break even. No business breaks even every day. But for every week that a business exceeds its break even, the CEO or General Manager lost no ground in his or her fight to grow.

The universal aspect of an hour is also addressed by the **Intrinsic Product Diskette** & its Private Label Agreement at the conclusion of the book. It allows the entire process to be disseminated throughout your organization by the right to reproduce the 20 different uses of the Appendix Diskette Programs of Performance Cash in their entirety. The Diskette also acts as an Appendix for the text of the author's first book, *How to Predict Year End Cash*.

I was browsing in the financial section of a bookstore to see if they stocked my first book *How to Predict Year-End Cash & Energize Any Size Business*. A gentleman asked me if I could recommend **one book** to help him understand financial reporting as an investor with no accounting background. As an experienced accountant & until **Performance Cash**, I had to say that book had **not yet** been written. Toward the end of completing the book a recent article from Accounting Today reminded me of that meeting of several years ago. Quotes from the article by two prominent accountants confirmed the situation; they testified before the Senate Banking Sub-Committee that the investor as well as the business manager are starving for knowledge & the profession is in the 1930's as far as financial reporting to the public.

There was a suggestion at the end of the Accounting Today article, "at the very least a Corporation should report daily sales", a natural start to helping the investor. But many stockholders in the equity market are business people first; they should be able to digest other simple data applicable to any business & to finally understand more about the financial reporting of their Investment. As a reader, remove the notion from your mind that *Performance Cash* is "accounting" in nature. The data should turn-on any business manager to what makes a business tick & wipe out the mental-block given to all numbers. **Four Benchmarks** are absolutely necessary to run any business. Think of the benchmarks as basics that a 16th century printer always knew about his business before the first CPA was born.

Business Control in a Nutshell

Benchmark #1--Sales per Hour, reveals the **Selling Price** for every hour **Earned from Unit Sales.**

Retail would be in the $25 to $60 range, **Manufacturing** $40 to $100, **a Professional or skilled workforce** would be $80 plus. The hourly rates are from my limited exposure. The uniqueness of the rate--- it can be at the same rate for start-up & mature businesses.

Bx #2--Labor Cost per Paid Hour, reveals **the Volume** of **Units Sold** (the basis of the Program copyright).

Retail would be in the $8 to $15 range, **Manufacturing,** $12 to $20. **A Professional or skilled workforce $20 plus.** The ranges would be much higher for start-up businesses when the labor force is programmable but in excess of that necessary for that phase of a business.

Bx #3--Performance Cash per Paid Hour must be **positive unless it is a start-up business. When positive the range will depend upon the Company's maturity in years less the extent it has re-invested cash flow in future growth.**

All Benchmarks are relative to anticipated hourly rates; when compared to Prior Results they are Truth versus Truth.

Performance Cash--Generic Definition: Established Companies produce positive Performance Cash each week--**Seasonal Companies** have positive & negative weeks with a yearly positive—**New Companies** have more negatives than positives with probably a year's negative--**Losing Companies** may occasionally produce a positive with probably a year's negative.

Performance Cash as Defined by an Accountant for an Accountant

Cash Sales + Credit Sales less Cash Out for all Expenditures except Income Tax, Capital Investment & Debt Service

There are **three** principal Drivers for Performance Cash---**Sales, Material Cost & Gross Payroll.** When the three are under constant scrutiny it means that the amount of Net Performance Cash always makes sense. There will be times when not much can be done about the Weekly results but I would not expect that to happen until a long way into the use of **Performance Cash.**

Sales Dollars through **unit** fluctuations in **price and/or volume** fall directly to the bottom line of **Net Performance Cash.** When **Unit Sales** deviate from last week or that anticipated the impact on Net PC is always known in **dollars** by the **average price** & **volume of** each principal product from day to day or week to week. Every business should know if it had a good or bad week through fact & that fact is **Performance Cash.**

Material Cost are generally the most significant & consistent expenditure as a percent of sales. Such costs can be anticipated & then tracked by deviations to the percent.

Labor Cost are more volatile than Material Cost because the effectiveness of gross payroll paid each week will always depend upon the number of units sold. **Net PC** subjects the control of **all personnel** for any business to that of Direct Labor as currently practiced by Manufacturing Facilities.

In Summary using Accounting Terminology

Weekly Performance Cash is how Sales, Hours & Cash Flow are exposed from all transactions of an entire business. ***Earned Hours©*** enables Performance Cash, traditional Cash Flow & Operating Earnings to be initially accountable to 100% of any forecast of unit sales, then subsequently to the **actual hours paid** as a weekly growth indicator through **Four Benchmarks** that can be compared to any other Company.

Shareholder Message Contained in an Open Letter

The book contains an open letter to Stockholders & Analysts that makes a case for why both Accountants & Analysts are barking up the wrong tree in both the preparation & reading of Financial Statements. With recent reporting irregularities, you as Shareholders seem to be the only offended party & maybe the only willing participant that can force immediate change to a long neglected & sad state for an Informed Capitalistic Society. The Epilogue of the Book tells you what you can do about the situation.

Dedication

I always considered Accounting as numbers speaking out about Performance with no one significant amount standing alone, but each in relation to another number. I am horrified by the present day deception in Financial Reporting; with greed taking the place of creativity. I do thank all facets of the Accounting Profession for their lack of curiosity as exhibited by constant resistance to my creative change. The result was a more comprehensive expression of Performance Cash as a measurable absolute for all Businesses.

Table of Contents

Appendix Table of Contents

Earned Hours© the Intrinsic Product
Diskette & Appendix to the
Power of the Hour

nn

Chapter One

100% of What?

Seeking Common Ground For and With Management

Operating Earnings is analogous to the final score of any professional sport's contest. But with sports, the story why one team won and another lost is in the next day's box score (or other form of statistics) of the game. When the box score is read there will be certain statistics indicative of the winner. Sure, there will be fluke wins that belie the stats. But good performance reflected with convincing statistics will still reveal that a team is ready and able for the next game or next month. Sporting Publications as well as fans know that weekly-summarized statistics will bring to bare why some teams are ahead of others in the standings. See page 6 for statistics at work.

As opposed to professional sports a 50% winning average is good in the business world. It will be like being in first place, because a winner will be exceeding a previous month's performance or plan, more often than not. The key to the sports analogy—**four benchmarks report sales efficiency the day after**—a day's—a week's—or a month's event. Management has more need than a sports fan for current sales data—his or her lively hood or investment is a stake.

In addition to playing on the same field as management, Accounting desperately needs help in support of Earnings per Share **(the all time benchmark)**. My collection of headlines and articles are below. I am not trying to replace EPS but to supplement it.

Why Controllers aren't making an Impact?—Strategic Finance April 1999

Why CFOs Get Fired—CFO Magazine April 2000

Why CEOs Fail—Fortune June 21 1999

Will Accounting Education Survive in the Future?—Management Accounting Quarterly Fall 2000. This was authored by a task force composed of three Accounting Associations and representatives from the Big Five Firms.

Wanted: Accounting model fit for the "new economy"—Accounting Today August 21 2000

The Fortune 500 Issue, July 2001 by Baruch Lev and Thomas A. Stewart

"accounting gives rotten information about the value and performance of modern, knowledge-intensive companies"

My reply to all of the above, "Accountants are taught how to deal with the presentation of business numbers, as opposed to certain numbers being the means to gain knowledge of any particular business". There is no common ground for an accounting student or a professional to gain knowledge of how a business survives and grows—*by selling each product of a business at the right price for the manpower employed.*

Special Report Strategies for the Accounting Professional—Practical Accounting July 2001. "There are 45,000 CPA Firms in this country of which 90 percent of the fees were from essentially doing the same thing" "The Firm of the future will free accountants to devote more time to strategic tasks like financial analysis and business consulting—for higher revenue services." The article spoke of "on line" and other "live service", deliverable instantly. *My reply—not if it is rehashing the same stale data that is available today.*

Newsday July 1, 2001—What's Cooking, Susan Harrigan

"restatements because of irregularities or errors averaged 155 per year 1998 to 2000 vs. an average of 49 from 1990 to 1997" She went on to describe the tricks to the manipulation of numbers. A Congressman's answer is to double or triple the SEC enforcement staff. ***Sounds like more aggravation for Industry.***

I believe established Corporations would welcome the Truth of my objective benchmarks, removing the pressure of EPS growth when the company is truly growing in other respects. Likewise accountants would have an **audit step** geared to operations. It is ironic that auditors concentrate on the Balance Sheet and devote little relative time to the principal interest of Equity Shareholders, **Operating Earnings.**—That alone must change if EPS is going to be the **driving force behind everyday shareholder value.**

Operating Earnings per Share gets all of the Shareholder's attention, was it always meant for such glory? A few pennies here or there impact Corporate Value in the millions. It is only one line of a Consolidated Statement of Income from a Company's many different entities throughout the world. That one amount from the last line categorizes and puts in order, from a myriad of sources, billions of transactions for the financial world to observe performance. Even when the rules are understood, those same financial people have a difficult time agreeing and expressing why reported earnings couldn't be matched with any prior actual or anticipated results. **Most variations to a *principle* have valid explanations.** Earnings per Share can not be the ***principal*** benchmark for operating performance unless it gets some desperately needed help in explaining deviations.

Accountants realize the shortcomings, but they have not adequately admitted that fact to the Public Investor. The 2000 Florida election for the presidency is a case in point. Voting has always appeared to the **public** as a clean process where numbers are objective and the results conclusive. The EPS process is tainted with certain weaknesses for a use and time not originally designed. My opinion—"Public and Private Accountants are not working together to enhance EPS and are almost going in opposite direction in their quest to find other supporting numbers for the *most sensitive principle to Equity Value*".

Accountants must seize the common ground when discussing a business with **Management before and after** Financial Statements are prepared. It is only a question of setting up a simple spreadsheet that sets in motion the proper framework of the past to impact the future. Just like **different general ledger software** is the basis for Different Presentations of Financial Statements, **the Intrinsic Product Diskette is the tool for the Truth within a Financial Statement and the basis on which the future should be planned.** Dealing in the future are where the real client fees are, especially when you give **Management** the machinery for a clear view of what may reasonably become its cash reality.

Soon after Accountants introduce management to the output of the four benchmarks, it will not be a question of what went right or wrong with the Operating Earnings of a Financial Statement, but a question of an Accountant joining with General Management to make things better.

	A	B	C	D	E	F	G	H	I
1	The Truth (winning) of Baseball has been available going back to the 19th Century.								
2	by daily operating benchmarks of **pitching** (earned run averages) & **batting** averages,								
3	comparable to the **Intrinsic components** of any business, **sales, hours & cash flow.**								
4				American League					
5			Baseball Team Winning Percentage						
6									
7	complete 2001 season			pitching	hitting	hitting		hit/pitch	
8		winning		earned run	batting	minus		place in	percentage
9	Team	percentage		average	average	pitching		standings	standings
10	Seattle	0.716		0.355	0.289	-0.066		1	1
11	Oakland	0.630		0.359	0.263	-0.096		2	2
12	New York	0.594		0.404	0.268	-0.136		3	3
13	Cleveland	0.562		0.465	0.279	-0.186		8	4
14	Minnesota	0.525		0.450	0.271	-0.179		7	5
15	Chicago	0.512		0.453	0.267	-0.186		8	6
16	Boston	0.509		0.417	0.265	-0.152		4	7
17	Toronto	0.494		0.431	0.263	-0.168		6	8
18	Anaheim	0.463		0.417	0.261	-0.156		5	9
19	Texas	0.451		0.575	0.276	-0.299		14	10
20	Detroit	0.407		0.498	0.261	-0.237		12	11
21	K. C.	0.401		0.490	0.265	-0.225		11	12
22	Baltimore	0.391		0.467	0.248	-0.219		10	13
23	Tampa Bay	0.383		0.498	0.258	-0.240		13	14
24	comparable to benchmark #1				comparable to Benchmark #2				
25	sales per hour				the labor cost per hour				
26	the price of each unit				the volume of all units				
27	**sales efficiency**				**labor efficiency**				
28				comparable to benchmark # 4					
29				(better not be negative)					
30				**growth efficiency**					
31			when combine with benchmark # 3, Performance Cash						
32									
33	Also going back to the 19th century before Impressionists Artists took their rightful place								
34	in the Art World of the 20th Century, was a **poet's** *(quote)* adapted by this author.								
35									
36	"Through four benchmarks Martin D'Amico sees a business bathing with truth *(light)*								
37	from thousands of irregular transactions *(touches)* , which at a distance, give a business								
38	*(the Work)* life. Where Accounting *(the Academic)* sees only a business in conformity								
39	to standards *(set down in regular separate positions).*								

No stats are conclusive. Three points regarding baseball stats versus benchmarksfour. Point 1-- Predicting Performance Cash will be more dependable for decisions because large numbers dominate averages as a year unfolds. Point 2-- daily sales performance need only average 100%. Point 3---

unlike sports, there is no other team to beat, only your forecast.

Imagine you as a Business Manager or Shareholder getting Percent Fluctuation data **every Monday morning** that will be **consistent with Quarterly Earnings per Share when reported.** I am not talking about Dollars as a Percent of Sales. I am speaking of data in the form of **Hourly Rates,** based upon the **Actual Hours Paid** to all employees and the **Actual Units Shipped;** both in four benchmarks format that will reveal the following.

The Percent Fluctuation in Unit Prices verses those anticipated.

The Percent Fluctuation of Labor Cost verses that anticipated.

The Percent Fluctuation of Overhead Spent verses that anticipated.

The Percent Fluctuation of Cash Flow Generated verses that anticipated.

I have extensively exposed **100%** to three of the Most Prominent Accounting Associations (AICPA, IMA & the NSA). Summarizing their responses:

"Not Invented Here", NIH

Accountants have been accumulating published data going back to the 1930s and the best they have come up with is "percent of sales" for **Operations** and **Balance Sheet** ratios.

Management and Investors must take the bull by the horn and demand more than they are presently receiving from accountants. In the long run, association members will be winners in the Earning Game by enhancing their fees as the best advisors for moving benchmarks in the right direction.

*"Civilization as we know it today, owes its existence to the engineers. These are the men who, down long centuries have learned **to exploit the properties of matter** and the sources of power **for the benefit of mankind"(1).***

Accountants have **not exploited the properties of data;** aside from SALES, EPS and PERCENT OF SALES, when have you ever read that **OPERATING numbers** from a Certified Financial Statement **benefiting** another business? Chances are never, because no two businesses account for **each transaction** and **accounting theory** in the same manner.

*"**Of the inventors**, only a fraction have the courage, stubbornness and energy to keep on bettering their **inventions** until they really work and to keep on promoting them until they persuade others to take them up"(2).*

100% was from chapter one of my first book started in 1995, I plan to see 100% and its four benchmarks through the fog of Operating Statements for the benefit of the Financial World.

(1) *The Ancient Engineers by L.Sprague de Camp, page 1.*

(2) *The Ancient Engineers by L.Sprague de Camp, page 5*

EARNED HOURS™ (the Program & Diskette) is the key to 100%

Earned Hours can be used to measure any volume of dollars versus a unit forecast. Units can be customers, transactions, product, square feet or chargeable hours. The equation for 100% efficiency of all labor cost with labor hours follows.

1000 units @$4.00 = <u>(sales of $4,000 less $640 in labor cost)</u>
$$\text{80 hours}$$

Sales $4,000/80 for the current week is $50.00 per hour, Benchmark #1

Labor Cost $640/80 = $8.00 per hour, Benchmark #2

100% = <u>1,000 units</u> or 80 / (1x1000) = .08 Hours
$$\text{80}$$

Proof, 1,000 units x .08 the constant per unit,

= 80 hours, the current week's goal in hours earned, hours paid and units sold.

Below, Earned Hours are being used to measure another "Volume of Dollars" those of Cash Flow, Benchmark # 3. The benchmark is still measuring the cash flow dollars based upon "unit sales".

If the Net Operating Cash Flow (without capital investment) after deducting all other expenses (excluding labor) of the business were $18.00 per Hour and $200,000 represented the estimated cash flow for an entire year, $200,000/$18 or 11,111 of people hours @ $8.00 per hour would represent an 100% efficient workforce.

Proof 100%   11,111 hours x $50 per hour = $555,550 in sales

11,111 hours x $8 per hour = $88,888 in labor costs or 16%

$8.00 / $50.00 = 16%.

If a seasonal retail or manufacturing facility only expected to be 30% efficient on Mondays or for the Third Quarter of the year, $8.00 / 30%, or $26.67 would be the Cost per Earned Hour for each one of those periods. The facility could only pay 30% x 80 or 24 hours to be 100% efficient for Mondays or any day during the Third Quarter.

Proof—24 hours x \$26.67= \$640, the original labor cost assumption for 100% of units shipped.

100% can be off in a distant future year and each day of a current year can have its own yearly percent milestone or Benchmark to operate against. Effectively comparing each current day with a Model or Long Range Plan of the future.

One principal product (blouses for a specialty retail shop, hamburgers for fast food, dinners for a restaurant, visitors for an entire theme park or web site) can drive the sales and mix of all products with a sales rate per labor hour that includes blouse (i.e.) sales plus all other dollar sales **divided only by blouse units. This means only the principal product, whether a manufactured product or a retail product, needs tracking on a day to day basis to determine 100% efficiency.**

A sole proprietor who draws no dollars out of the business would use 2,000 hours at \$1 per hour. **The hours are merely the measuring device for 100% efficiency in sales units.**

The key .08, need not be accurate at the start, it only needs a good guess, before long the actual benchmarks will give you the Truth for going forward.

Manufacturing facilities have been using "labor earned" versus "labor paid" for years.

Since Earned Hours includes all labor cost at any facility it is easily aligned with Activity Base Costing.

The trend today is toward **multi-product** retail outlets such a Wal-Mart, where employees and their hours are no longer serving one product, making it impossible to determine product-profits. **The profits will be determined by how well or efficient all the people hours, serve all the products.** Likewise, software people serving visitors to their web sites.

Bxxxx and its Earned Hours (tm) (if a Corporation so chooses) **is the statistical connection that allows every estimate, budget, plan and model of Cash Flow and Operating Earnings including EPS, to be accountable to unit sales, whether units are transactions or a multitude of customers buying different products.**

Earned Hours is bridging the gap between anticipated and actual results to make a Financial Statement more understandable and accountable to an **Investor-Reader.** Not only does it do it for the **Investor**, it allows **Management** to talk more consistently and intelligently about their Company.

100% Cash Flow Defined for the Benefit of an Accountant

As an Accountant, you are aware Operating Earnings are subject to accruals and inventory accounting. My definition of **Operating Cash Flow** includes Credit Sales as a Cash Positive along with all other cash outlays, but does exclude Capital Assets because **the weekly accumulation of Earned Hours (tm) was devised to be connected to the most consistent business indicator possible (Performance Cash).** Nevertheless the Weekly **Performance Cash** and/or **Operating Cash Flow,** as defined is easily reconcilable by an accountant to: **Earnings, (before and AFTAX) or to EPS, thus allowing Earned Hours(tm) to retain its integrity and accountability to unit sales**.

It is also Management's choice to break out any **Performance Cash and/or Earning Forecast** of the Corporation into "unit or market segments", "facility segments" or combine an entire Company into a "one unit" business. The one unit can be customers (transactions), hamburgers, dinners, theme park visitors, web site visitors, square feet, pounds, etc.

Referring to page 16 the two most important elements of the weekly cash flow are **the selling price** of each **unit &** the **total number** of **units** sold. Two benchmarks from **100%** can capture that data on a weekly basis. **Benchmark #1,** which is called **Sales per Earned Hour**, captures the actual selling price of all unit sales on an average basis whether selling the same product to many purchasers or selling many different products to a few customers.

Benchmark #2, which is called **the Labor Cost per Earned Hour** captures the total number of units sold. The reason it is labeled as **"labor cost",** it measures the effectiveness of an entire labor force (as budgeted or planned), to sell a certain volume of units (as budgeted or planned). Although the benchmark is measuring unit sales by the deviation to the dollar & cent benchmark, the variance is measuring the efficiency of the **payroll dollars** as contained in a budget or plan.

Benchmark # 3—Performance Cash—assumes credit sales are collected; which is the only anomaly compared to traditional Operating Cash Flow (before capital expenditures). It is also the true cash return on each hour of labor paid. **When a Corporation tracks year to date Performance & it is growing as an hourly rate & such cash flow is maintaining expectations, management can be assured that favorable EPS will eventually follow, regardless of the earnings indicated by monthly & quarterly results.**

Benchmark #3 Weekly Performance Cash starts as a **forecast ($18.00),** & is referenced by a **certain** number of units being shipped. If the actual "units shipped" volume were at 100%, **Bx #2** & its Earned Hours would be on target ($8.00) as the volume productivity indicator. You & I know that **hitting a 100% "year to date" unit volume** projection is unrealistic. Fear not, the deviation **(-$.56)** to the dollar & cent amount of **benchmark #3 (performance cash)** will be itemized as to cause in dollars & cents per hour: **as to selling price (#1+$.74), as to labor cost, (#2 - $.30) & to all other expenses as a residual (#3 - $1.00).** Although 100% "year to date" may remain unrealistic, each week's plus or minus **($.17)** of efficiency will be the key to everyday management decisions.

E a r n e d   H o u r s™

are the means
to expose Three Principal Components
from a Financial Statement

1   product & the selling price of each
2   people & the number of units they sell
3   Performance Cash from the product & by the people
merely
B e n c h m a r k (reference) the Components
with Four Hourly Rates & Make Any
Estimate - Forecast - Budget - LR Plan - Model
Accountable to 100% of the Units Shipped
as often as daily or weekly

	99%		100%
		Per Hour	
B e n c h m a r k	**per unit**	**actual**	**assumed**
#1 sales	$51.25	$50.74	$50.00
#2 labor	$8.38	$8.30	$8.00
#4	$42.87	$42.44	$42.00
#3 Per Cash	$17.61	$17.44	$18.00
cost of 1% inefficiency		$0.17	
($. 56) *detail*		$17.61	
+ $. 74 *sales*	(less units @ higher price per unit)		
($. 30) *labor*	(overtime)		
($ 1 . 00) *all other*	*spending*		

Left for Management

*Improve Bx #3 & Bx #4 from Week to Week
or Year to Year
& from financially justified Capital Expenditures*

The Intrinsic Product

A tangible product may have many components that make up the finished product to a customer and drives the *sales* of the product to all the businesses that handle the product from manufacturing through distribution. But the **tangible product is not** the intrinsic product that drives the ***profits*** of any individual business that handles the product.

The *profits* of individual companies who handle the product, will tend to rely more on how *efficient* the entire labor force of the organization is, when carrying out all the duties from manufacturing to distribution of the product. Efficiency is an element of time. Especially when all the **hours of all personnel, expressed as a unit of ONE, is the value that can be reported upon and control the profit plus cash of a business.** At different stages of a product's journey through a facility employees will be adding value whether they are directly handling, supervising others, packaging, designing, or distributing the product. The time it takes for each particular product to go through a plant or retail facility is **Intrinsic** to its cost and sales value. That time, as a decimal equivalent of an hour is the unit's rate, to be *earned* when completely manufactured or sold.

When a production unit is complete or sold the budgeted hourly decimal portion of the product becomes "Earned" When many units are sold the accumulations are ***Earned Hours*™**, a productivity measurement.

My basic theory is simple; all employees of a business determine *profits* of a business more than the products themselves. Employees work hours, when paid they become paid hours, which can now be compared to earned hours.

The Diskette

Earned Hours is bridging the gap between anticipated and actual results to make a Financial Statement more understandable and accountable to an Investor-Reader. When accomplishing that end for the Investor, it allows Management to talk more consistently and intelligently about their Company.

The Intrinsic Product in the Form of Earned Hours ©

Presents Uses

Budgets, Long Range Planning & Modeling becomes One Yearly Process

Start-UP Ventures have a Daily Milestone from Day One.

Seasonal Businesses can have "day of the week" & different weekly Benchmark #2s

Adapts to an easily applied Pricing Formula

Backlog analysis for Performance Cash Forecasting

Other Marketing Cuts of Product Line

Specific Material Margin analysis of All Products

Chapter Two

Product Businesses

The truth of a business lies within the components of each transaction. The **3 Ps** & most revealing components of each transaction are **Product, People,** & the **Performance Cash.** The effectiveness of financial reporting is how **sales, hours** & **cash** flow are **exposed** from all the transactions of a business. Bxxxx exposes two truths of an entire business. **A) The cash flow of each unit sold B) The cash flow of each hour paid.**

Although the **Bxxxx** process appears to be a comparison to a budget, that is not the case. It is merely a means to **Product** Sales, **People** Efficiency, & **Performance Cash,** the basics of any business.

Have you ever wondered why a Model, a Long-Range Plan, a Budget & for that matter, **every day operating results**, never have common source data for comparison to each other? **Modeling or planning with Bxxxx will do away with annual budgets & replace them with one comparison process to any & all-future years.**

This chapter of the book was organized to guide any person with an interest in any business, through the process by creating four benchmarks from a **single transaction.** After educating a reader with its **every day** use the number of transactions are expanded for planning & controlling an entire business.

There are many different ways to present management performance. One of them, **Operating Earnings,** lacks cash detail & an everyday format, with marketing generally ignored. **Bxxxx and Performance Cash done weekly or daily** can eventually **negate monthly closings.** An accountant can still reconcile any or all benchmarks to operating earnings monthly, quarterly or yearly.

"Earned Hours"© transforms all **unit** sales to **hours** & is the **basis for the truth & uniformity in enumerating Sales, Labor Cost & Cash Flow consistently from: one month to another, from one company to another, from anticipated results to actual results.** When used to supplement the Operating Earnings of a Financial Statement it is analogous to Windows & its ability to find files & unravel data from a PC. Neither the information nor the benchmarks are subject to modification at a latter date when Financial Statements are prepared.

Chapter Contents

When the truth is different from that anticipated, & that will always be the case. You will have two courses of action—make operating adjustments to get back on track or modify the anticipated results by means of the benchmarks.

The Transaction

My book *How to Predict Year End Cash* developed two benchmarks from a year's paid hours and reduced them to a unit of one, the **sales per hour** and the **cost per hour**. The sales per hour measures **sales efficiency (selling the right units at the right price)** while the cost per hour measures **labor efficiency (total unit volume of sales for the manpower employed).**

Very few companies know the efficiency of their sales effort or their entire workforce, but even so, actual data **after the fact** rarely results in any cash impact. Therefore, before a year starts the two rates are formulated from four simple assumptions. A week's example: **(1)** 1000 customers will be sold **(2)** @ $4.00 each **(3) during** 80 hours of labor **(4)** @ $8 per hour.

The sales per hour $4,000/80 = $50 per hour is Benchmark #1.

The cost per hour as stated, $8 per hour is Benchmark #2

All four assumptions will naturally fluctuate during the year, but the two actual deviations versus the above benchmarks will **always** expose the truth.

Here is the key. Every time a customer is sold during the year it will earn **.08 hours** (80 hours /1000 units). A customer is now represented by the hourly rate, and many customers are represented by an **accumulation of earned hours.** The day or week's goal in units is 1000 customers and in **earned hours** is 80.

To continue the example, the actual results for any given week, 990 customers were sold at $4.10 each equaling **$4,059.** Actual sales of 990 units earned .08 each or 79.2 hours, $4,059/79.2 = **$51.25 per earned hour.** The mix and/or price improved $1.25 per hour and $.10 per customer. Both extensions will give the same total, $99 for the week.

23

During the same week, clerks stayed some extra hours and were paid $24 overtime or **$664** in total. After a division by 79.2 earned hours, the result is **$8.38 cost per earned hour. Inefficient** by $.08 per hour, because 990 customers were sold vs. 1000 budgeted. But,,, cash flow is favorable $99 less $24 for overtime ($.30 per hour and part of the $.38 unfavorable rate per hour).

Earning Hours is nothing more than attaching a predetermined rate in hours to units or customers prior to actually selling them and then measuring how good the unit sale assumptions are after the fact.

The Truth

	B	C	D	E	F	G	H	I
1	The Truth						assumption	actual
2		Sales Dollar Variance Verse Assumed					input	input
3			Benchmark #1				page 12	page 12
4		Actual Sales			$ 4,059	customers	1000	990
5		Assumed		$4,000		sell price	$4.00	$ 4.10
6		Flexed		($40.00)	$ 3,960			
7			Variance		$ 99	labor hours	80	80
8	linked	990	x	$ 0.10	$ 99	labor cost	$8.00	$8.30
9								
10							Earned Hours	
11	proof	$ 4,059	990	=	$4.10	to be earned	0.0800	
12						earned		79.2
13		Sales Unit Variance Verse Assumed					Hours Paid	
14			Benchmark #2					80.0
15		990	1000		99%			
16			inefficient		1%			
17		$8.00	0.01		$0.08			
18								
19								
20								
21								
22								
23					Per Hour			
24					A	B	if the assumed	
25					earned	paid	were a forecast	
26	Sales		Bx #1	$ 4,059	$51.25	$50.74		$50.00
27	Labor Cost		Bx #2	$ 664	$8.38	$8.30		$8.00
28	Value Added		Bx #4	$ 3,395	$42.87	$42.44		$42.00
29	type any number in f44 only			$ 2,000				$ 25.00
30	Cash Flow**		Bx #3	$ 1,395	$17.61	$17.44		$17.00
31								
32	A--Truth based on units shipped							
33	B--Truth based on hours paid							
34	If $17.61 vs. $17.44 were YTD results, the difference,							
35	of $.17 is the total cost per hour of being 1% inefficient							
36								
37	**check book change + or - change to accounts receivable							
38								
39		Acct Rec	proof		Cash			
40	Beg Bal	$ 20,000			$ 2,000			
41	sales	$ 4,059						
42	collected	$ 3,000			$ 3,000			
43	payroll				$ 664			
44	* all other				$ 2,000	$ 2,000	for linkage from e29	
45	End Bal	$ 21,059			$ 2,336			
46	net change	$ 1,059			$ 336			
47					$ 1,059			
48	performance cash (as above)				$1,395	by the unit & hour		
49								
50								
51								
52								

	A	B	C	D	E	F	G	H	I
1	Reconciling &			Introducing Performance Cash					
2	Monthly Operating Earnings verse My Weekly Cash Flow								
3									
4	Results of Monthly Closings				Results of Weekly Cash Flow from prior chart				
5	Sales				Sales		Bx # 1	$50.74	
6									
7	Material Cost				Material Cost		% of Sales	n/a	
8									
9	Labor Cost				All Salary & Wages		Bx # 2	$8.30	
10									
11	Manufacturing Expense								
12					One Accumulation				
13	Inventory Accounting Plus or Minus								
14									
15	*Manufacturing or Gross Margin*				**Acct. Rec. change*		+ or -	**	
16									
17	Selling & Admin Labor								
18					*Performance Cash		Bx # 3	$17.44	
19	Selling & Admin Expense				*My Operating Cash Flow before*				
20					*Capital Expenditures*				
21	Accrual Accounting Plus oMinus				*Debt Service*				
22					*Income Tax*				
23	*Operating Earnings before Income Tax*								
24									
25	Inventory Accounting opposite of above + or -								
26									
27	Accrual Accounting opposite of above + or -								
28									
29	Depreciation Included Above, Plus								
30					Sales less Labor		Bx # 4	$42.44	
31	*Operating Cash Flow before*								
32	*Capital Expenditures*								
33	*Debt Service*								
34	*Income Tax*								
35									
36					** the only rule, a credit sale is a cash				
37					positive on the day of the sale.				
38									
39									

The Expansion

	A	B	C	D	E	F	G	H
1	The Expansion		(see data change below)					
2								
3	a larger workforce over a more extensive product line							
4								
5								
6		Labor Connection, Basic				no fringes		
7			head count	hours	hourly rate	(000)		
8		Gen Mgr	1	2,000		$ 80		
9		salary	4	8,000		$ 120		
10		hourly		40,000	$12.00	$ 480		
11		totals		50,000		$ 680		
12		cost per earned hour			$13.60	Benchmark #2		
13								
14								
15	Sales/Labor Connection							
16				(000)	sales			
17		yearly	price per	yearly	hour	% hour	earned	labor cost
18		units	unit	sales	allocation	allocation	per unit	per unit
19	hamburgers	100,000	$ 4.00	$ 400	23%	11,628	0.1163	$ 1.58
20	sandwiches	50,000	$ 5.00	$ 250	15%	7,267	0.1453	$ 1.98
21	platters	30,000	$ 7.00	$ 210	12%	6,105	0.2035	$ 2.77
22	salads	20,000	$ 3.00	$ 60	3%	1,744	0.0872	$ 1.19
23	coffee	200,000	$ 1.00	$ 200	12%	5,814	0.0291	$ 0.40
24	soda	400,000	$ 1.50	$ 600	35%	17,442	0.0436	$ 0.59
25		800,000		$ 1,720	100%	50,000		
26								
27	other sales			$ 80			the key	
28		totals		$ 1,800				
29	hours to be earned			50,000				
30	Sales per Earned Hour			$ 36.00	Benchmark #1			

The Expanded Truth

	A	B	C	D	E	F	G	H
1	The Expanded Truth without Cash Flow							
2				the key	A Week's Actual Performance			
3								
4				the proof				
5	yearly		hours	for year	dollars			earned
6	units	hours per	extended	$ per	extended	wks units	week's $	hours
7	100000	0.1163	11628	$ 1.58	$ 158	1,900	7500	220.9
8	50000	0.1453	7267	$ 1.98	$ 99	1,200	5900	174.4
9	30000	0.2035	6105	$ 2.77	$ 83	550	3900	111.9
10	20000	0.0872	1744	$ 1.19	$ 24	400	1230	34.9
11	200000	0.0291	5814	$ 0.40	$ 79	3,800	3600	110.5
12	400000	0.0436	17442	$ 0.59	$ 237	8,000	11000	348.8
13	800000	hours	50000		$ 680	15,850	33130	1001.5
14						other	1700	
15						sales $	$34,830	
16						payroll $	$13,100	
17								
18					$34,830/1001.5			
19								
20	Results of Weekly Cash Flow							
21	Sales		Bx # 1	$ 34.78		$13,100/1001.5		
22								
23	Material Cost		% of Sales	**				
24								
25	All Salary & Wages		Bx # 2	$ 13.08				
26								
27	One Accumulation			**				
28								
29	Performance Cash		Bx # 3	**				
30								
31						Value Added per Hour		
32	Sales less labor		Bx # 4	$ 21.70 ———▶		Benchmark # 4		
33								
34	** not available at this time							

Sensitivity Test

	A	B	C	D	E	F	G	H
1	The Expanded Truth without Cash Flow					(without hamburger sales)		
2				the key	A Week's Actual Performance			
3								
4				the proof				
5	yearly		hours	for year	dollars			earned
6	units	hours per	extended	$ per	extended	wks units	week's $	hours
7	100000	0.1163	11628	$ 1.58	$ 158	0	0	
8	50000	0.1453	7267	$ 1.98	$ 99	1,200	5900	174.4
9	30000	0.2035	6105	$ 2.77	$ 83	550	3900	111.9
10	20000	0.0872	1744	$ 1.19	$ 24	400	1230	34.9
11	200000	0.0291	5814	$ 0.40	$ 79	3,800	3600	110.5
12	400000	0.0436	17442	$ 0.59	$ 237	8,000	11000	348.8
13	800000	hours	50000		$ 680	13,950	25630	780.5
14						other	1700	
15						sales $	$ 27,330	
16						payroll $		$ 13,100
17								
18								
19								
20	Results of Weekly Cash Flow							
21	Sales		Bx # 1	$ 35.01	$27,330/780.5			
22					price & mix sensitive			
23	Material Cost		% of Sales	**				
24								
25	All Salary & Wages		Bx # 2	$ 16.78		$13,100/780.5		
26						unit volume sensitive		
27	One Accumulation			**				
28								
29	Performance Cash		Bx # 3	**				
30								
31						Value Added per Hour		
32	Sales less labor		Bx # 4	$ 18.23	→	Benchmark # 4		
33								
34	** not available at this time							

Martin D'Amico

Model Transactions

*Using the same approach with different **unit and hours** assumptions for the **last year of a Model** (in this case the 3rd year).*

The Business Plan—the Company is a Hotel with sales expected to reach **$12,000,000 in year 3**. Benchmark # 1 is to be **$48.00** per hour in year 3 to accommodate the new personnel requirements of the entire Corporation, **250,000 hours.**

The personnel needs consist of a complete hotel staff. Such a Corporate composite will amount to **$2,500,000** with an average wage of **$10.00** per hour or **250,000 hours.**

New Assumptions subtracted from sales during the last year of the Model.

Material Cost will be at 25% for the life of the model and in year 3, $3,000,000.

All other operating expenses, including all fringe benefits but excluding fixed asset purchases and depreciation will amount to $4,000,000 in year 3.

Year 3 must be back phased to year 1.

All operating expenses are estimated to be $4,450,000 in year 1.

Cash Flow Model—Benchmark #3—Year 3—$2,500,000

Cash Flow Model—Benchmark #3—Year 1—$ 50,000

Year 3 $10.00 per Hour   #1 Sales—$48.00   #2 Labor—$10.00

Year 2   not shown

Year 1 $.40 per hour   #1 Sales—$48.00   #2 Labor—depends on

payroll dollars

See the plan above reduced to a weekly performance cash breakeven for the first week of operations. (XL #7)

Playing with the Truth

A CFO is a necessity for every business. They are my call letters for the Central Financial Force of a business. Not every business can afford the salary of an expensive Chief Financial Officer, but the Force must still be there. The Financial Force can be in the person of a Chief Financial Officer, Controller, an Accountant, or yes, a General Manager.

While numbers in the form of the four benchmarks give overall cash flow direction and control to a General Manager, the following is an aspect of the "Operating Earnings Game" and Inventory Accounting that should be understood to some degree.

Even when the actual **sales dollars** are equal to those planned for any period in time, they never seem to generate the operating earnings anticipated. An accountant begins to interject fixed and variable costs, expense timing, **shipments out of inventory**, absorbed overhead, etc.etc. If you as a manager would have asked why earnings were better, when sales are less than those expected, which you rarely do, you would have received a more confident tone. But, with the same set of reasons. The principal reason for the bulk of any earning variation compared to anticipated results is—**production and sales data are rarely the same on any given day, week, month or year; inventory is either increased or depleted**

For a **Manufacturer** there is no way to predict a meaningful variation between production output verses sales output (Actual Sales) in any one week, month or quarter. Predicting an inventory change at year-end may be more successful because the impact may be of little significance compared to an entire year's output.

Inventory Accounting, accounts for swings in output and can influence operating earnings during any particular time frame with the impact beneficial or detrimental to earnings in any one period. When a benefit, a manager must discount the earnings effect in his mind since it can't be counted upon in the future and it certainly can't be spent as cash in any way. Inventory build-ups must also be discounted when peering into the future for long range planning purposes.

Controlling your business weekly through four benchmarks without Inventory Accounting will be controversial, but give it a try. Let your accountant worry about Inventory, he created it along with its shrink. Consider all purchases of material as just another expense, controlled as a percent of sales, inventory taking can be limited to unit counts without the assignment of dollars.

That is the bad news about manufacturing. The good surrounding a standard cost system; an element of time has already been set up for all products. You merely scale up proportionately, each product's hours to account for the 50,000 hours to be paid, to arrive at the hours you expect to earn when sold. If your company doesn't have a cost system you will be creating one based on all employees, instead of those only working in the manufacturing facility.

You will get more financial information from weekly reporting of benchmarks and their cash flow, why play the Monthly Earnings' Game.

Growing the Truth

Investment Bankers have always had it right. On a yearly basis they take unit sales of a basic product along with selling prices as cash income and deduct all the cash outlays or cost for getting to and then servicing the market and call it *cash flow from operations.* Simple enough for every General Manager to understand both the data and the task at hand.

Until accountants get a hold of the data. **Monthly** they break the product into *dollar segments* put units into **inventory as dollars** where they *check out the efficiency* of producing the units. Then they take the units out of inventory as dollars, ship the **dollars** to customers and **break up** the cost **into great detail** by cost of sales and departments. After which accountants **add accruals** and reflect on all the pluses and minuses as one figure called **operating earnings. Monthly** they take operating earnings in total then add or subtract what went into inventory and accruals to **reconcile** to the net cash flow.

<u>**Yearly, accountants take the detail of the operating earning dollars along with all those inventory changes and accruals and use that as the document for planning several years out.**</u>

All the **underlined items** are no longer necessary while the *bold remain* part of *Performance Cash and the Bxxxx process.*

As you learned on prior pages, Benchmarks use the forecasts of units and the prices of each **from the last year** (your choice 2 to 5 years) of a business Model or Plan. The benchmarks convert the units to hours in that **final year** and connect them to sales dollars and cash flow and call it 100% efficiency. **100%** (of the units shipped) is back phased and linked by hours and units into each year of a plan. **In effect, every day or week of a current year is being compared to the last year of a Model or Long Rang Plan with the deviation to the benchmark, the variance to the current year as well as the last year.** Any year that exceeds 100% efficiency and cash flow per hour simultaneously, has achieved the Company's growth potential. **Because, the Company lost no ground to cash flow while increasing unit volume.**

To grow in an organized manner as opposed in an unorganized manner, will save a great deal of cash and aggravation over the long run.

35

When discussing business with even remotely interested parties, I have learned to ask one pertinent question, "Are you interested in growing your business". If not, the business part of the conversation is brought to an immediate end. There are managers and business owners who are very satisfied with the income of their business as a hobby or otherwise, and don't want to put any more energy into their endeavor.

To discuss growth, I want to return to an equation from my book, ***How to Predict Year-End Cash & Energize Any Size Business.***

Cash + Data + Energy = Growth—Data is the "How To" of predicting cash.

The Energy is that of a General Manager.

Have you ever notice how many executives quit good positions and become consultants. They get burned out because each assignment has the same or even more pressure than the past. Yet, they sense or know the urgency really isn't there, and "number one priorities" have no end in sight. Also, when companies lay off by the masses. What were all those people doing that a company can survive without? They were probably accumulating useless data and planning how to make it better.

The expending of **excess** energy or cash has to be planned if one dollar **outside of normal operations** is going to be spent and starts to drain any cash. Whether paid to an employee or an outside vendor, tasks outside of normal operations need an order or priority because there never seems to be enough cash or time to go around. The difficulty in filling requests for something not planned, multiplies when the original cash of a plan is falling short, and makes no sense to expend. When there is no plan, all non-operating tasks should be abandoned; otherwise it is called unorganized planning with no inter-departmental coordination.

If business planning connotes success why not plan. If nothing else, there will be business satisfaction when something planned several years back is achieved. Without a plan, a manager will never know if he could have done better when actual results are recorded. Even though you believe that you have been successful without planning you have no point of reference; you will never know true success, only that you reacted very well to prevailing conditions as they were encountered.

Achieving an operating cash flow of $1 million before tax in the third year after the current year ends, is a plan. In support, I assume you have the underlying unit sales and the price of each major product line and the tasks to back up the plan. If you already have a positive cash flow and cash is rising from year to year your business appears to be in good shape. The increase in cash indicates growth although the rate of growth may not be consistent with the industry's or the past. If less than the industry's and you are satisfied, good, but maybe you could grow faster and share more cash with employees who may not to be satisfied.

If a long-range plan is going from a negative flow year to a positive out in the future, decide which **year** is the turning point to the plus side. That swing year is critical to planning because up until that time cash is leaving the business and not finding its way into your pocket. You will never know which month, nor can anyone tell you, the exact turning point because there are too many variables to such a formula. The critical issue, getting there during the year, not the exact month. There is a way you will know that you have arrived at the turning point. Make the adjustments to Earned Hours by means of the two Benchmark Rates per Hour in the following year of the plan and verify that the net flow remains positive for an entire year.

The Benchmarks of the business will tell management where to exert its energy for growth, while winning the daily earning game will be a matter of moving the Benchmarks in the right direction.

To capsulate all of the above with only two rules

Premise: **Cash is the bearer of growth**; in the order of significance the following are the **only six elements for growing cash.**

more customers or units—better pricing of units

lower labor costs—lower material costs

lower overhead cost—lower working capital

The most significant external force of a business, **customer demand** is intertwined; not with the ability of an accurate forecast but with the **mere ability to** put **a unit forecast on paper.**

I can't tell you how to manage the external force of customer demand, which is what your industry experience is bringing to the business table. The principal internal force, overhead spending item five, is where accountants concentrates their efforts in both recording and advice. My advice, after a budget is in place for a current year, it is very difficult to put programs in place that would favorable move the external force of a business in that year.

Rule One: Allow overhead spending for improving the external force of a business when documented and connected to an exact dollar and cent movement to benchmark # 3, cash flow per hour. You as general manager must approve the expenditure and then adjust the cash impact in the current long-range plan for all future years.

Rule Two: With the current overhead budget in place there will be no unfavorable spending unless there is an overhead offset in another expense category or in another department.

To sum it all up, a business must spend money to get money just as you spent money to buy this book. That spirit to spend is there when a business first starts-up because an entrepreneur realizes that he or she needs customers and even borrows money to see that they get them. That spirit begins to wane, as many businesses mature and pay down debt with **seemingly no new cash** needs for the business. The successful companies never let the spirit disappear whether money is to be borrowed or generated through operations.

Modeling/Planning by

Hours & Benchmarks

	A	B	C	D	E	F	G
2	Model by 50 week year in (000)						XL# 5
3							
4							
5			a year is	50	weeks		
6	ABC Inc.		In Benchmark Format				
7	Hours to be Earned			369767	739535	1800000	3000000
8			partial year /				
9			start up $$	year 1	year 2	year 3	year 4
10	net sales			$ 10,850	$ 25,200	$ 54,000	$ 90,000
11		Bx 1		$ 29.34	$ 34.08	$ 30.00	$ 30.00
12	other variables						
13							
14	cost of materials		$ 50	$ 901	$ 2,092	$ 4,482	$ 7,500
15							
16	all personnel costs		$ 50	$ 8,250	$ 15,000	$ 30,000	$ 40,000
17		Bx 2		$ 22.31	$ 20.28	$ 16.67	$ 13.33
18	fringes on all labor			$ 2,063	$ 3,750	$ 7,500	$ 10,000
19							
20	operating expenses		$ 50	$ 4,635	$ 6,000	$ 8,020	$ 10,000
21	work capital		$ 50				
22	Op. Cash Flow			$ (4,998)	$ (1,642)	$ 3,998	$ 22,500
23	Earnings, Per Cash	Bx 3		$ (13.52)	$ (2.22)	$ 2.22	$ 7.50
24							
25	capital expenditures		$ 200	$ 100	$ 100	$ 100	$ 100
26	equipment leases						
27	debt service						
28							
29	Cash Flow b4tax		$ (400)	$ (5,098)	$ (1,742)	$ 3,898	$ 22,400
30							
31	Accumulative Cash Flow		$ (400)	$ (5,498)	$ (7,240)	$ (3,342)	$ 19,059
32							
33	deprecation			$ 30	$ 40	$ 50	$ 60
34							
35	Op. Earn with Depr			$ (5,028)	$ (1,682)	$ 3,948	$ 22,440

41

Chapter Three

Professional and/or Service Business

The truth of a business lies within the components of each transaction. The **3 Ps** and most revealing components of each transaction are **Product, People,** and the **Production of Cash.** The effectiveness of financial reporting is how **sales, hours** and **cash** flow are **exposed** from all the transactions of a business. My four benchmarks are references that expose **The Truth** of an entire business.

This chapter of the book was organized to first guide, any person with an interest in any business, through the process by **creating four benchmarks from a single transaction.** After educating a user with its **every day** use the number of transactions are expanded for **controlling and planning a complete Professional or Service Business.**

A Professional Business is one whose principal resource, the hours of the workforce, are supplying a service to clients or customers and not primarily selling a tangible product.

The profits and **Cash Flow of a Professional Business** will rely on how efficient the entire workforce is when carrying out all the duties of supplying its service to clients or customers. Efficiency is an element of time. Especially when all the hours of all personnel can be expressed as a dollar and cent rate. **When an hour of service is rendered it becomes "earned" and it can now be compared to how much that hour sold for and cost.**

Although the case material in this text follows that of a Law Firm, the last part of this edition shows how to apply the concept for any Service Company.

Chapter Contents

Because the **Four Benchmarks** are readily available from three (sales, payroll, and hours) **operating numbers, it requires no change to any accounting systems currently in use.**

If you are reading this chapter without the benefit of completely understanding chapter 1 & 2; mentally divide your service business into two distinct segments. A purchasing segment that can predict, estimate, budget, plan & control **material cost & all subcontracting** costs as a **percent of sales.** Segment Two is the Service End of the Business & the basis on which this chapter was written. When you physically apply Bxxxx to your own business, an **Accountant must separate Sales & Material plus Subcontracting Costs with their own accounts.**

Earned Hours are Chargeable Hours

The backbone and truth of your Professional Business is how many hours: you **charge,** the **rate billed** for each of those hours and your **labor cost**. Nothing you don't already know. Benchmarks will roll those variables into how much cash is generated by those daily activities exposing the true output of your entire business entity. Merely give those chargeable hours a new name—**Earned Hours.**

The Transaction

My book *How to Predict Year End Cash* **developed two benchmarks from a year's paid hours and reduced them to a unit of one, the** sales per hour **and the** cost per hour. **The sales per hour measures** sales efficiency (selling the right units at the right price) **while the cost per hour measures** labor efficiency (total unit volume of sales for the manpower employed).

Very few companies know the efficiency of their sales effort for their entire workforce, but even so, actual data **after the fact** rarely results in any cash impact. Therefore, before a year starts the two rates are formulated from three simple assumptions. **A week's example: (1) 600 client hours will be sold (2) @ $166.67 each 3) from a workforce whose weekly payroll is $32,000.**

The Sales per Hour as stated or $166.67 per hour is Benchmark #1-- - $100,000/600

The Cost per Hour $32,000/600 = $53.33 per hour is Benchmark #2

The week's actual billings or sales were $115,000 with a payroll of $29,500 and there were 550 hours charged by all the employees of the firm. The net or the two summary transactions was $85,500 before other expenses, **page 60 (d20).**

The single most important element of your cash flow is the service hours or fees billed. Like a manufacturer the most important component of efficiency is a relationship between payroll dollars paid out (without fringes). Unfortunately and also similar to a manufacturer the week's production or hours "to be charged" are rarely the same as those billed for the same period of time. Thus there are two "sets" of efficiency needed to determine the weekly cash flow of a business. Over a year's time, the proximity of the **"two sets of efficiency"** to each other will determine the total profit as well as the cash flow of a Professional Service Business. The "invoiced" verse "production" hours may come close together by year end, but you still must know the weekly impact if you are to do anything about any glitches that may arise in managing your people and the cash performance of your business. At year's end is too late to discover that you billed 30,000 hours and had 35,000 as charged indicated by hour bookkeeping from employee accountability. Unlike present day manufacturers with limited direct labor efficiency standards, the output of Four Benchmarks (Bxxxx) is based upon all employees that may work at a facility.

Bxxxx gives a user the ability to monitor **production verse sales.** You should try to invoice clients each week, a practice that can only help cash flow. The amount can be partial billing or the hours of a complete contract, what ever is more practical and it need not be aligned with the work performed during the same week. Retainer billing need not wait to month's end; in fact it is better done in four weekly cycles broken down by the alphabet. A month is too long a time to monitor employees or the cash performance of a business. **Weekly** cash performance based on the condition that invoices represent cash flow is the most consistent and manageable period of any business. Even if salary people are paid semi-monthly, it is much easier to convert their hours to a week and conform them to the hours of your wage employees.

In case you didn't notice in the above paragraph, I bent accounting terminology when I refer to hours billed as a positive weekly "cash in" item. No operating reporting system can start without the success of each week's billing. The collection of the resulting accounts receivables will determine the effectiveness of an accounting department, but billing must be the leading efficiency factor for operating and now part of my definition of operating cash flow. Let your accountant reconcile the weekly benchmarks of cash flow when preparing his financial statement; you also can easily observe the simple arithmetic on the bottom of the **Truth Page (page 60).**

Grasp the benchmark process by first extending the numbers of the above transaction to yearly figures by a factor of 50 weeks, page 58. They now read as follows along with the inclusion of $2,000,000 as operating expenses.

Sales, page 58, d28 $5,000,000

Labor Cost, page 58, d30 $1,600,000

Operating Expenses (includes all fringe benefits, d31)

$2,000,000

Operating Cash Flow (without Capital Expenditures, d32)
$1,400,000

The Cash Flow per hour $1,400,000/ (600 X 50 weeks) or 30,000 hours =

$ 46.67---Benchmark # 3, page 58 e32

$ 53.33---Benchmark # 2, page 58 e30 labor, remains the same

$166.67---Benchmark # 1, page 58 e29 sales, remains the same

$113.33---Benchmark #4, is introduced, sales less labor, page 58 e31

Reiterating, the payroll to support 30,000 of billing hours is expected to be $1,600,000 **(page 58 f9)**. The **efficiency** of the entire workforce as indicated by the benchmarks is based on **payroll dollars.** Hours actually paid will be a factor to determine what I refer to as yield. Yield is-- **the chargeable hours verse all hours paid.** The case assumes 30,000/54,000 or 56% **(page 58 d12)**. The yield indicates that some personnel will be called upon to do overhead type functions that can't be specifically passed on to a client. Yields as well as the benchmarks are based upon each salary person working 40 hours in a week and 2000 hours a year. This should put a stop to all Professional Service Firms playing the chargeable hour game by making professional people account for every hour of their year. It will also eliminate a great deal of unnecessary "hour bookkeeping" by just keeping track of **only** the hours charged along with a weekly yield factor. Yes, one person can have a yield in excess of 100%; it is the overall yield of the firm that is important for weekly cash performance.

There is a third element of time that is important. Those are the hours that each person works in a year. A partner may put in 3000 hour a year but may only charge 500, while a clerk 2500 with 2400 chargeable, and there are other strictly overhead people on salary that may put in 2500 hours for 2,000 paid, and none are chargeable.

As a third element of time, **capacity utilization** is now chosen as that term. Chargeable hours are still the numerator with the denominator all the hours people are **on the job.** The capacity utilization denominator from the data immediately above would be 3,000 + 2500 + 2,500. Both the yield and capacity percents are important and should be tracked when new people are added to the workforce. Maintaining the percent's interval between both sets of hours is the most important contributor to your cash profit. Yes all employees must submit the total hours on the job but do not have to account for their non-chargeable hours.

The expansion of the labor connection on **page 58 (top grid)** is merely a headcount of all **salary (d6 & d7)** people with the dollars of regular pay. Regardless of how many hours they may work during the coming year they are pre-programmed to work 2,000 hour a year and 40 hours a week. **Wages (d8)** or hourly people need only be taken as a group along with their hours and the estimated total dollars. That is it, your manpower or manning table is that easy.

To grow as a Professional Business beyond a "one professional" enterprise, a sliding fee scale per hour must be part of the expansion to accommodate overhead personnel that can not command income comparable to the Professional Principal. This is necessary to remain competitive as you grow and allow the firm to unburden itself with routine work that can best be done by a subordinate at a lower rate. The management of those overhead hours and the perception of the scaled serviceable rate as charged will instill client confidence that he or she is being charged a fair rate for all service. On a contract basis the mere fact that you have more than one rate will impress a client that they are being fairly charged even without an itemization of the contract hours by each category of the scale.

The last item of labor, or the hours to be charged, **(page 58 f25),** can not be ascertained until you make some kind of weekly or yearly projection of your **Fee Schedule** and the mix of your different rates. Only make a good estimated guess because it won't be long before you have actual data that can be brought on stream any time during a year. Your mix of rates on an actual basis is the single most important element to weekly cash flow volatility. For **page 58 line19, Court Room** time on the same chart, I am not saying that only partners will get $500 an hour for entering a courtroom. It is the average of all people who will be charged out for being in a courtroom. I am also not asking you to change your billing rates for each service; I am merely trying to determine the true mix and its cash impact.

On **page 58** your yearly fees are proportionate to the whole as a percent **d25 & e25.** The percent of hours are divided by the units to establish the hourly rate **(Column G line 19)** for each class of service. You now have the basis for comparing the actual mix of daily or weekly production. From that mix of chargeable hours on those time slips, you can compare the effectiveness of your payroll dollars, **benchmark #2, $53.33 (page 58 e30).**

When hours are billed for the week on **page 59 (the Truth without Cash Flow)** from its own set of invoiced hours, please observe the truth of your business **d21 & d25.** The billing of **$115,000 (g15)** is bringing in **$209.09 (d21)** per hour while your workforce cost are **$53.64 (d25)** per hour; not favorable at 99% vs. **f25.** Notice **e12** on the chart, 550 hours charged is far less than the hours billed for the week, **page 48.** Please remember that the 600 hours are those billed during a particular week, and may or may not include those actually charged in the same week, or even the same month if you wish to continue billing with that cycle. Rarely will production hours (550) be equal to those invoiced during a week. There is no need to know the amount of hours billed each week, the important fact are the dollars billed year to date and its relationship to production hours.

Examine the **Cash Flow Truth page 60 f6;** $15,000 is the normal dollar variance to a sales budget while line **19 columns G & H** states the efficiency of your payroll dollars, **$53.64 f19,** actual verse **$53.33 h19,** or that anticipated.

Move to **Column G line 22, $104.55 vs. $46.67, (h22)----the true cash flow based on fees invoiced.** Made up of Column E, cash in from a good billing week less **Column F $53.64,** the outflow of payroll dollars based on each production hour of the week. In summary line 22 **Column G** the Cash Flow per Hour of **$104.55 (g22)** compares very favorably to the **h22, Benchmark #3 $46.67.** In case you are interested and you need not be, in the accountant's behalf **line 32 through 40** has the reconciliation to actual cash that considers the difference between accounts receivable collected and the sales for any period in time.

Why was **$104.55 Bx #3** so good for the week? A good mix of Courtroom time at the highest rate must have been in the hours invoiced. Only Court Time **billings** and "other sales" are capable of such a cash impact and surpassing the forecasted weekly sales of $100,000. Good

Courtroom **production** took place during the week upholding the efficiency at 600 hours. Please note that it is very difficult to present specific numbers in an example that will bring home a convincing point to impress a reader with the Truth of his or her own business. Yes it was a good week in every respect, but the important point I want to make, the **$104.55 g22,** rate may be very difficult to maintain for the entire year. The averages will be hard to move as the year unfolds. Even if you choose to stay with your monthly billing cycle, the three production criteria are worth repeating.

Production The Earned Hours for the Week

Yield   Earned Hours/Hours Paid

Capacity   Earned Hours/Hours Worked

None of the three include dollars, yet you will know more about your business than any closing of the books will accomplish. If you are going to keep track of the three, why not attach them to the dollars or the three Ps of a business, **Product, People** and **Performance Cash and weekly you will have everything necessary to predict your cash for the year.**

It is more important to send invoices weekly than close the books monthly to control and predict your cash flow.

Now for the accountant and if you are curious, **(page 60 d22) $57,500 of Cash Flow.** Each week can be a plug to agree with the checkbook change from one week to the next, plus or minus the Accounts Receivable change. Capital Expenditures are not included in my definition of Operating Cash Flow and would have to be excluded.

Nothing above need wait for a yearly budget process; it can be started next week, with the truth of the business coming forth in a matter of a few weeks.

Other Service Businesses, page 61 worksheet

Selling **hours** for other **Service and Entertainment Businesses** is no different from above, whether an **Architect, Accountant, Interior Designer, Contractor, Maintenance Firm, Theme Park** etc.

The **fourth worksheet, page 61** of this chapter, **Customers or Transactions as Units vs. Chargeable Hours** presents a choice how a Law Firm or any one of the above businesses could proceed. The right portion of the sheet is the same data as the first three sheets of the chapter, with a consolidation of the product line into only two rates ($500 for Courtroom Time and All Other at $100) tying into the annual sales of $5 million, **page 58** . Of course Benchmarks #1 and #2, Sales and Labor, **i5** and **i10** agree at **$166.67** and **$53.33 with all prior pages 58 to 60.**

The choice is whether client hours should be classified as a **"customer or transaction unit"** as opposed to an **"hour unit"**. The number on the left, 106,250 **(a9)** units could easily be visits to a Theme Park i.e. Disney, with a one customer or transaction rate of $47.06 (not shown or $5 mil/106,250). The 106,250 customers could be **bank transactions, web site transactions, square feet, lineal feet** for contractors, **cubic feet** for a moving company, etc, etc. If desirable 106,250 units could be broken down into products and selling price ranges (**a7 & a8** of the example only chose two). The validity of Bx #1 at **$92.59 (d10)** per hour and Bx #2 at **$29.63 (d5)** are different from the right because the hours are those to be paid vs. those to be charged.

Emphasizing the point, the Case numbers on the left make a further cut of the customer market (or transaction) or unit count into two products. This would be ideal for a **Hotel & Restaurant** combination business. Any business can use the **total customer or transaction** approach as a starting point and later branch off into extending the product line into two or more products.

There is no difference to the left and right approaches, although left will be useful for comparison to other firms. The right is not comparable because it doesn't include the total hours of the entire workforce within the hour base. The total hours are easily ascertainable if ever evaluating another firm being considered as an acquisition.

If hours charged only have a couple of different rates or if the service for the entire business is the same I recommend the "customer or transaction" approach, This is especially recommended if there is no time lag (Doctor's Office) between extending the service and the billing (today's work is tomorrow billing) with very little contractual service. If there is a blend of the two, "tomorrow vs. contractual billing" merely set up a "two product" product chart similar to the left. If there is an extensive range of both employee skills and their billable rates as a law or accounting firm, along with a great deal of additional overhead, go for the right.

Other benefits of Earned Hours

The use of earned hours allows you to look forward with greater certainly as you estimate, quote (or bid) and book contracts of the future. Always convert the total dollars of each to hours. This need not be in great detail and the divisor can even be an average rate. Whatever hours are accumulated per contract should be placed in three buckets: **possible, probable and then into backlog when a contract is signed.** Constantly be aware of the hours in each of the three buckets. Those **hours to be earned** will give you an idea of the manpower needs for the near future. When your contract plate is full you can quote prices with more profit and visa versa when business is not holding up. It will not be long before you will always be thinking of your business by the earned hour for both the past and the future. What isn't earned can be used as a vacation, you have earned those hours too, but now you will have more cash in your pocket.

Conversion to Earned Hours

	A	B	C	D	E	F	G	H
1	The Expansion & Conversion to Hours Earned					XI #2A		page 58
2			(see data change below)			only change the blue		
3	a large professional with a sliding fee structure							
4			Labor Connection, Basic			no fringes		
5			head count	hours	hourly rate	(000)		
6		Partners	8	16,000		$ 910		
7		salary	10	20,000		$ 440		
8		hourly		18,000		$ 250		
9		totals		54,000		$ 1,600	doesn't have to be	
10	adjustment to Earned Hours			(24,000)			exactly the same as d30	
11		cost per earned hour		30,000	$ 53.33	Benchmark #2		
12			yield	56%				
13	before the key is applied (hours/units), all the hours must be allocated							
14	by their respective sales dollars, without other sales $$'s (D25)							
15	Sales/Labor Connection							
16				(000)	sales		don't touch	
17			yearly	price per	yearly	hour	%/hour	earned	labor cost
18			hours	unit/hr	sales	allocation	allocation	per unit	per unit
19	Court Room	5000	500	$ 2,500	52%	15544	3.11	$ 166
20	Senior Fees	6000	150	$ 900	19%	5596	0.93	$ 50
21	Junior Fees	3000	125	$ 375	8%	2332	0.78	$ 41
22	Prof Service	10000	75	$ 750	16%	4663	0.47	$ 25
23	Clerical Ser.	6000	50	$ 300	6%	1865	0.31	$ 17
24							the key	
25		30000		$ 4,825	100%	30000		
26								
27	other sales			$ 175				
28		totals		$ 5,000	30000			$ 5,000
29	Sales per Earned Hour			$ 166.67	Benchmark #1		
30		labor cost		$ 1,600	$ 53.33	Benchmark #2		$ (1,600)
31		operating expenses		$ 2,000			Benchmark #4	$ 113.33
32	Performance Cash			$ 1,400	$ 46.67	Benchmark #3		

Benchmarks #1, 2 & 4

	A	B	C	D	E	F	G	H	I
1	The Expanded Truth Without Cash Flow					EX #2B		page 59	
2									
3				the key's					
4				proof	A Week's Actual Performance				
5	yearly		hours	dollars	Produced or Charged		Billed		
6	units	hours per	extended	extended	hours i.e.		week's $		
7	5,000	3.11	15544	Court	109		hour or $$		
8	6,000	0.93	5586	Senior	90		detail		
9	3,000	0.78	2332	Junior	130		not		
10	10,000	0.47	4663	Prof Ser	140		necessary		
11	6,000	0.31	1865	Clerk Ser	81		for billing		
12			don't touch		550		of g15		
13	30000	hours	30000		550				
14				page 25	550	detail must agree with e13			
15	from the text--- change						$ 115,000	billing imputed	
16	for the effect after you test					imputed	payroll $	$ 29,500	
17							change both to your heart's content		
18					$115,000		$29,500		
19					divided by	550.0	divided by	550.0	
20	Results of Weekly Cash Flow								
21	Sales		Bx # 1	$209.09	versus	$ 166.67	125%		
22						budget			
23	Material Cost		% of Sales	**					
24									
25	All Salary & Wages		Bx #2	$ 53.64	versus	$ 53.33	99%		
26						budget			
27	One Accumulation			**					
28									
29	Operating Cash Flow		Bx #3	**					
30									
31						Value Added per Hour			
32	Sales less labor		Bx #4	$ 155.45		Benchmark # 4			
33					versus	$ -			
34	** not available at this time					budget			
35						#4 can be just as Truthful if			
36						don't want cash flow data			

The Cash Flow Truth

	A	B	C	D	E	F	G	H
1		The Truth with Cash Flow				XL #2C		page 60
2			Sales Dollar Variance Verse Assumed					
3				Benchmark #1				
4		week	Actual Sales			$115,000		
5		week	Assumed	text		$100,000		
6				Variance		$15,000		
7								
8								
9								
10								
11								
12								
13								
14					Sales	Production	Combined	budget
15					Per Hour			per week
16					A	B		/50
17					550.0	550.0		
18	Sales		Bx #1	$ 115,000	$209.09		$209.09	$ 166.67
19	Labor Cost		Bx #2	$ 29,500		53.64	$53.64	$ 53.33
20	Value Added		Bx #4	$ 85,500			$155.45	$113.33
21	actual, from below first			$ 28,000			$50.91	$ -
22	Cash Flow**		Bx #3	$ 57,500	Performance Cash		$104.55	$ 46.67
23								
24	A--Truth based on hours sold							
25	B--Truth based on hours charged &/or earned & workforce pay							
26	Both Truths should come closer together in Dollar Value as the year unfolds							
27	If not, there is a glitch in your hour accountability.							
28								
29	**check book change + or - opposite change							
30				to accounts receivable				
31		Acct Rec	proof		Cash			
32	Beg Bal	$700,000			$20,000			
33	sales	$115,000						
34	collected	$127,000			$127,000			
35	payroll				$29,500			
36	* enter here first				$28,000	28,000	repeat for linkage to d21	
37	End Bal	$688,000			$89,500			
38	net change	$12,000			$69,500			
39					$12,000			
40	Performance Cash (as above)				$57,500	by hours sold		

The Unit of Your Business

	A	B	C	D	E	F	G	H	I	J
1				Units verse	Chargeable hours			XL #2D	page 61	
2					for the same business					
3	Labor	hours 30,000 @$53.33 per hours = $1,600,000 for Black & Blue								
4				labor per hour					labor per hour	
5	$1.6mil/54,000 =			$29.63	Bx #2	$1.6mil/30,000=		$53.33	Bx #2	
6	Customers Profile	(000)	hours	EH	Hours Profile		(000)	hours	EH	
7	50,000	$55	$2,750	29700	0.59	5,000	$500	$2,500	18000	3.000
8	58,250	$40	$2,250	24300	0.43	25,000	$100	$2,500	18000	0.600
9	108,250		$5,000	54000		30,000		$5,000	3000	
10	sales per hour Bx #1			$92.59		sales per hour Bx#1		$166.67		
11			The above amounts are for a year, below for a week.							
12	Actual Customers, one week 1/50 of above					Chargeable Hours, one week 1/50 of above				
13	1000	$55	$55,000	0.59	594.0	100	$500	$50,000	3.000	300.0
14	1125	$40	$45,000	0.43	486.0	500	$100	$50,000	0.600	300.0
15			$100,000		1080.0			$100,000		600.0
16	sales per hour Bx#1				$92.59	sales per hour bx#1				$166.67
17										
18	Actual Customers, one week @80%of units					Chargeable Hours, one week @80%of units				
19	500	$55	$27,500	0.59	297.0	50	$500	$25,000	3.000	150.0
20	563	$40	$22,520	0.43	243.2	250	$100	$25,000	0.600	150.0
21			$50,020		540.2			$50,000		300.0
22	paid 1,081 @$29.60=$32,000,Bx #2				$ 59.24			Actual Bx#2		$ 106.67
23				budget	$29.60				budget	$53.33
24	Actual Bx #1		$92.59		50%	Actual Bx #1		$166.67		50%
25										

61

Value Added per Hour, Benchmark #4

A benchmark to measure the growth of a company from period to period.

When any financial statement is read, aside from **Earnings as a Percent,** or as **"Earnings per Share"** there is no way to ascertain whether a Company is truly growing from one period to the next. Since these two references are laden with Accruals, Inventory Accounting, various Income Tax Rates and "interpreted" Accounting Theory along with a Management bias, both **Earning Rates** are in need of enhancement. You say you rely on the Cash Flow Statement for true growth. You are a rare breed of Accountant or Investment Banker. Operating Cash Flow is a very elusive number that can only be abstracted by an experienced professional through a certain collection of numbers.

Benchmark #4 is very simply, benchmark #1 less #2. Only a company just starting out with heavy financial backing could absorb payroll dollars exceeding sales. Unlike normal Bx #2 all **Fringe Benefits** are added to payroll dollars and a "net" part of Bx #4. Normal Bx #2 is pure "cash flow out" in the form of **gross payroll**. It does not include any accruals for employee benefits, only the share of employee benefits withheld from their pay. I have been very careful not to subject benchmark #2 to accounting theory, but be straight "weekly cash out." When fringe benefits are added through payments or accruals at the end of any month, quarter or year, they should be a fairly consistent rate as a percent of labor cost and can even be a gauge to the validity of the amount accrued in any one period.

When **benchmark # 4 is growing as a rate per hour**, it means that each hour "paid out". is being exceeded by sales dollars. For a service business, it states that the sales excess is improving relative to the corresponding labor cost per hour. Such excess is a positive trend from one period to another.

Regardless, whether a **Service Business** or a **Company** selling tangible products, **when benchmark #4 is increasing** along with **benchmark #3, cash flow per hour (before capital improvements),** the Company is growing.

After deducting capital expenditures (for a **different "net"** number of operating cash flow dollars verses the benchmarks) and that residual is not improving from one period to the next, employing **B#3 & B#4** to measure justifications may help the situation. **Both** will earmark the cash flow objective as, "a dollar and cent rate" prior to the capital expenditure's approval.

The next page is a format for benchmark #4 on a yearly basis for a Major Corporation. It can be part of the budget process, either as a directive to each operating unit or it can be the result of each unit's budget process as submitted.

When benchmark #4 is net of fringe benefits **including those of all Corporate Officers and Directors,** the rate per hour will indicate whether the top executives are taken too much relative to their performance.

	A	B	C	D	E	F	G	H
5			Value Added					
6			The Entire	Corporate	Man' f	Research	Retail	Software
7	millions		Corporation	Office	Plants	Center	Outlets	Service
8	net sales		$ 2,300		$ 445		$ 900	$ 955
9	labor cost with							
10	all fringe benefits		$ 453	$ 7.3	$ 82.4	$ 25.1	$ 103.7	$ 234.5
11	value added		$1,847.0	$ (7.3)	$ 362.6	$ (25.1)	$ 796.3	$ 720.5
12								
13	hours (000)		18,700	60	5,200	830	7,900	4,710
14								
15	sales per hour		$ 122.99	$ -	$ 85.58	$ -	$ 113.92	$ 202.76
16	labor Cost per Hour		$ 24.22		$ 15.85		$ 13.13	$ 49.79
17								
18	VA per Hour		$ 98.77		$ 69.73		$ 100.80	$ 152.97
19	Corporate & Research	Driver	18,700		18,700			
20	labor cost per hour			$ 0.39		$ 1.34		
21								
22								
23								
24								
25								
26								
27								
28								
29								
30								

Open Letter to Public Accountants

I have been in touch with many partners of all the remaining major Public Accounting Firms about Performance Cash, an exciting new weekly truth in reporting concept. It was not from lack of trying but none of my colleagues have been good enough to feedback or question. The same 100% concept which is the basis of Earned Hours© was presented in a variety of versions. I offer a partial excuse for their non-action--- "getting their house in order from a great deal of unprecedented bad press".

As this book is finalized for its publisher, I have arrived at an opinion for the lack of interest by Public Accountants. It is the same lack of interest given to budgets by auditors during an engagement and the root of some of their own Profession's problems. They are always examining and comparing results but not primarily to budgets. Whether such actual client results are quarterly or yearly, audit comparisons are generally made to a prior period. Earning results may even be consistent with an earlier quarter but still be filled with variances of all kinds, client subjectivity, and sometimes agendas that Public Accountants have no easy way to know or discount when evaluating the reasonableness of the Operating Performance from one time frame to the next. **Of primarily importance Public Accountants have no accurate means to relate unit sales variations and its impact on Operating Earnings.**

Significant Earning variations without a good feel for comparative unit sale references can not be measured for materiality or performance and must be left to Management comments that may or may not relate to reality or to units. The comments are general in nature because **the client themselves have no accurate means to relate units to earning dollars when they too examine their own variances within operating performance.** I believe the Public Accounting Profession is not cognizant that such a void exists within Industrial Reporting. Industry generally know price and volume as it relates to sale dollar results verse a budget, but is not aware of the exact mix impact as it relates to Cash Flow and Earning results. **Mix is the Earning Driver** whether measuring *actual to actual* or *actual to budget,* because some **Individual Products are more profitable than others regardless of price.**

The starting source for measuring performance as well as materiality is the Client's own current budget for an operating year. The budget may be the only documentation that establishes a link to:--- units with prices--- to sales dollars--- to profit or loss. Oh yes, there is a great deal of bias as well as undisclosed and various agendas by Management. Many budgets have their own inherent weaknesses, not flexible or current and are not always properly documented with exact product detail. But it is the best starting point to Segment or Division commitments.

Working in the depths of reporting Manufacturing Operations is not where today's CF0s come from, but is the best source for an MBA in

Business because a day's output before a sale even takes place is the prime source of efficiency data. Whether a manufacturing supervisor has a budget or not, he or she knows how many units they have to produce in each day or week or how many units represent 100% efficiency. Ultimate reporting exposure---is to first convert a dollar budget to 100% for the manufacturing floor and then to a selling budget somewhere between 95% to 105% of what manufacturing produces. Selling 100% of that manufactured is optimum for Financial Performance. There is no way to predict the exact variable output in a manufacturing budget verse a sales budget. But there is a way to know the deviations and the impact on earnings in a day, a week, a month, a quarter and a year.

---Actual Earned Hours© and its four benchmarks for Manufacturing—verse— Earned Hours© and its Four Benchmarks for Sales---**and the exact difference between the two.**

As an experienced auditor speaking from an ex-Controller's advantage point, Earnings of a Corporation are made up of all its operating parts or Divisions. Whether a Corporation has good or bad budgetary habits, each Division is usually committed to their annual budget. That forecast should be the principal source for testing and comparing results during any audit engagement. It is the best source of performance for unit sales variances verse those operating results anticipated. Wouldn't it be great data if each Division wrapped up weekly performance with one dollar and cent benchmark that indicated variances by unit volume, price and spending from a budget? Which in turn was summarized for an entire Corporation's own volume, price and spending sensitive benchmark? And that same Corporate dollar and cent benchmark could be expressed as a stock multiple with a distinct relationship to EPS.

And when both the CEO and CFO thought the original budget or forecast of the Corporation was no a longer a valid barometer, Management had documentation as to unit volume, price, and spending by Division that supports a new hourly benchmark producing another stock multiple.

Would it not be the nirvana of reporting for a Corporation to prepare a singular weekly benchmark report in dollars for Management and its Auditors that would eventually become the basis of reporting to the Investment Public and all other interested parties?

The Hourly Rate of Performance Cash vs. Operating Earnings as a Share Value Multiple

Performance Cash as a multiple to Shareholder Value is only meant to supplement Earning per Share at the outset. Until one gets comfortable with its singular use let me give you some assurance by first establishing PC and its exact relationship to Operating Earnings after Taxes are deducted. Foremost is my intension that Performance Cash should include extraordinary occurrences when they are truly expenditures (discontinued operations including its labor). As defined and allowed by the Accounting World, these occurrences are sometimes "one time deductions" to circumvent the penalizing of Ordinary EPS with the caption **After Extraordinary Items.**

A quick reconciliation of PC to Operating Earnings after Taxes---Performance Cash + or -

(1) Plus or minus Inventory change (if Material Inventory has increased it is subtracted along with COS Material in arriving at PC) originally included with Operating Earnings

2) Plus or minus non-cash items (accruals) originally included with OE

3) Less Depreciation + Amortization included with OE

4) Less Interest Expense included with OE

5) Less Income Taxes (after deducting Deprecation, Amortization, Interest,) **equals Ordinary OE and the basis for Earnings per Share.**

Of extreme importance, PC is easily reconcilable to OE by most every accountant; but **such an accounting is not necessary for a Shareholder.** The "four line items" (see the brief presentation of Performance Cash) deducted from sales as expenditures to equal PC were intended to be easily defined, uniform per industry, revealing and consistently present in any business with the least amount of accounting theory. Depending upon the Industry the four line items are easily expandable for other significant **operating expenditures.**

Item 1, Inventory Expenditures for raw materials, materials and ingredients as a cash outlay had to be part of PC derivation because Inventory Accounting is the Accountant's most subjective influence on Operating Earnings. As a deduction, PC is merely recognizing an inevitable expense before subjectivity rears its head. Besides, incurred Inventory additions will ultimately be deducted as an expense during the same year or in a subsequent year.

Item 2 (non cash) items are merely adjustments for the original exclusion of such journal entries not necessary when arriving at PC. Plus or Minus **Accrual changes made by accountants within OE** are meant to reflect current spending levels and future liabilities (pensions) being earned by employees and matched to their current output (sales and/or production). The Accounting Profession calls it **"matching cost with revenue"** and it is the principal culprit for adding to Reserves in good times. Management is always preparing for rainy quarters when a few pennies may be needed during a less than desirable future performance. The best reflection of current spending trends in reporting Operating Results is to use the amounts actually spent during the reporting year; that is what Performance Cash does.

3) Depreciation (a non-cash deduction not included in PC) is an expense based upon prior cash outlays for Fixed Assets when first acquired. Performance Cash does not capture such cash outlays unless classified as an expense. Major Fixed Asset purchases can be intentionally delayed and-- *Capitalization verse Expense*--is subject to individual Corporate Policy for a cut off in materiality (i.e. only items in excess of $5,000). Most companies should produce enough Performance Cash for such purchases each year but not each week. If not, more debt will be required. Any Corporation that wants to add cost justifiable plant or equipment to expand should be certain that the return on Performance Cash will increase as a rate per hour after the investment is in place.

Interest Expense, item 4 for any business is determined by each Corporation's capital structure. With no two Entities organized alike, Company to Company comparisons of interest cost are not relative to operations or Performance Cash. If a stockholder wants to evaluate whether his or her Investment is over extended with its debt, Interest should be added to the payments of Principal on the debt. In combination Interest and Principal payments are referred to as Debt Service. If PC is not covering its yearly Debt Service, a long term shareholder should bail out. A footnote to a balance sheet should be the ratio of Performance Cash to Debt Service.

Income Tax is the 5th reconciling item when going from PC to Aftax Earning dollars. If any business doesn't generate enough Performance Cash for the taxes expected to be paid, it too will run into a need for additional debt. But aside from that, here is a rainy day reserve that can always be swayed a few pennies here or there. Since Performance Cash precedes and is generally higher than Operating Earnings I wish I could suggest a percent yield or constant ratio to arrive at Operating Earnings. Impossible, because Income Tax which is generally the largest deduction in arriving at OE, contains more accounting theory than any other item of expense and would defeat the entire objectivity of PC as a performance monitor.

All of the above comments coincide with my definition of a mature business---**one that generates Positive Performance Cash over a year's time.** If a year's PC is positive one dollar it is possible to have Operating Earnings, but **if PC is substantially negative it is doubtful whether OE can be positive or meaningful for shareholder value.** A Start-Up Operation or non-mature business is so named because it is anticipating a negative PC for a certain period of time up to a few years. When there is a cross over to positive on a sustaining basis, the weekly amounts will be crucial to its future investment value. **But for every minus week, the Start-Up will need a positive week at least equal to the negative, additional capital, or an extension of its debt.**

The accumulation of **Positive Performance Cash Dollars without** the "Paid Hour" calculation, is the singular most revealing dollar indicator to Shareholder Value and the path used by Investment Bankers that evaluate any business for Purchase or Capital Reorganization. Operating Earnings or EPS is not the Banker's source for Investment Value as itemized by the five above reasons. Investment Bankers capture the Performance Cash of a business over the last several years and arrive at what they think has been its yearly average in dollars. They would then try to predict how much Performance Cash would be produced by unit sales and their respective prices in each future year and program a Value after considering any additional Fixed Assets needs. If the method is good enough for Investment Bankers it should be available to all Shareholders. Especially when Performance Cash can be reduced to the 3rd of Four Hourly Benchmarks on a Weekly Year to Date basis that is also **a return upon the most significant and current Investment in any Business, the People and the Hours they work.**

I am **not** saying---when Performance Cash and Aftax Operating Earnings are multiplied by the same number you will get equal Share Value. At the outset, a multiplier for EPS **($.74 x 16)** may have to be eight times that of PC **($5.30 per hour x 2)** in a similar economic environment ($11.00 of market value). But who knows? When the Investment Public grows to **understand and have assurance in the truth** and **objectivity** of **PC per Paid Hour** as an **Investment Equity Monitor,** the indicator may settle into

73

a commensurate multiplier equal to that of EPS (16 x $5.30 and $85.00 of value).

Due Diligence and the palm Controller

Although a CEO signs on to Earnings per Share, how binding and conclusive is it?

Although a CFO signs on to EPS he or she is really only the collector of numbers for the benefit of all the Officers, Directors & Shareholders of the Corporation.

Legislatures are naively holding CEOs & CFOs responsible for a Financial Statement. Consider this; the variety and multitude of each and every operating facility reporting in a Public Certified Statement, the mass of transactions, and all the Accounting theory subject to many people's interpretation at home and abroad. This is an unfair as well as an unrealistic approach to truth in Financial Reporting.

Once again, the accumulation of Operating Cash Flow is the "singular most revealing dollar indicator to Shareholder Value & the path used by Investment Bankers that evaluate any business for Purchase or Capital Re-Organization". If the method is good enough for Investment Bankers it should be available to a CEO Weekly & to Shareholders Periodically.

Operating Cash Flow is synonymous with the third Benchmark of Performance Cash and the pulse to relieve everyday Earning pressure from the Equity Marketplace recently imposed on CEOs and CFOs. In addition the Earning Multiplier of Due Diligence and the palm Controller is the Investment return upon the most costly and current cash outlay of a Business, the People and the Hours they work.

I am confident that CEOs and CFOs will understand several XL worksheets with imbedded cell   comments that are driven by customer demand as to the price, mix & volume of units as well as current spending levels at each location. The worksheet produces a summary mini profit and loss report with variances to that anticipated that records what actually happens at each Operating Facility along with the capabilities to up date an entire year as often as desired.

Investor bad taste & the disrespect that Accounting Professionals absolved themselves from was not helped when the SEC passed the responsibility buck to Management without a reliable mechanism to re-structure confidence. Due Diligence and the palm Controller is the **Equalizer for Shareholders** in language all interested parties will understand. It beats defending Accounting Theory & materiality to Governing Committees.

A brief presentation of Performance Cash

(Meaning to Act as a Summary to this Book)

A CEO's Due Diligence & the CFO's palm Controller,

what they Sign on to is what Shareholders & Analysts Get.

Pure Profit & Loss is readily Available from Every Business

Earned Hours© **enables Performance Cash, traditional Cash Flow & Operating Earnings to be initially accountable to 100% of any forecast of unit sales, then subsequently to the actual hours paid as a weekly growth indicator through a   Cash Flow benchmark (#3 of 4) & a Share Value Multiplier that can be compared to any other Corporation.**

Accounting sees a Business in conformity to standards; I see the truth of a business as a reflection of 3Ps: Product, People, & Performance Cash. The effectiveness of financial reporting is how Sales, Hours & Cash Flow are exposed from all transactions of a Business.

All interested parties can be availed common financial data in four hourly benchmarks; only Bx #1 & #3 are here in presented. There is no better abbreviated information for any person interested in the financial performance of a business & it is easily available weekly.

Many Executives have a tendency to get turned off by numbers in any form. Much too often profit expectations have been disappointing or have not materialized; they leave number crunching to Accountants while they retain their own seat of the pants control mechanism. **Performance Cash & its 100% process is that tool.** Users **need not know** how it is done, only that the process **can be done with a minimum of effort by their Accountant when directed or ordered to so.**

100% IS

a Plan, Budget or Estimate that sells 16,000 units / in 1000 hours that produces $6,500 of Cash Flow or $6.50 Per Hour (Benchmark #3)

OR

1000 / 100% (1) over (16,000) equals .0625 Earned Hours© for every Unit Sold

and

$6.50 in Cash Flow will be anticipated for every Actual Hour Paid.

Performance Cash takes the fog out of Profit & Loss reporting of a business for the benefit of non- accountants. **PC** is not skewed or camouflaged by accounting theory or subjective materiality & **must be sustained as a positive** prior to attaining any Profitable Earnings. **So why not report PC, in raw form verifiable by bank deposits & hours paid**

before Accountants transform the same data to their generally accepted & subjective standards?

I know of no other written material or process that has financially distinguished one successful business from another in a **Certified Statement of Operations**. No two companies have the same definition of materiality or handle all transactions in the same manner (i.e. Capital Acquisitions). I have taken what is available & consistent for every business & put it in the form of Four Hourly Benchmarks as well as a mini Profit & Loss statement. That is---**Cash Sales plus Credit Sales In,** minus **All Operating Cash Out,** excluding interest & capital expenditures. **It does lay a business bare in dollars for Management & in Benchmarks for Shareholders.**

The uniqueness of the **operating statement in benchmark format** is the **Earned Hour©** copyright that breaks down **dollars** of sales--**to units** of sales--, & then converts the units **to hours**---while at the same time presenting **cash flow** that can be **recorded, understood & projected as an hourly rate** in relationship to how many hours each employee is **paid.**

The strength of the process--- when there are deviations(always)--- to a plan, budget or a revised estimate, the **exact variances** are documented by **unit volume**, (how many), **unit price** (by major products), & **third, cash spending** in four major categories. They are 1) labor, 2) fringes on labor, 3) all other operating expenses & 4) as a lump sum, all material, raw material, ingredients & subcontracting costs. The expenses include **no accounting journal entries**; that is what true cash flow is all about.

Important--- Before PC came along, there was an accounting theory that has **never been accurately** put to practice, **Break Even---. Any forecast of PC is convertible to a Weekly Break Even point**. The Break Even of PC is primarily based upon different unit volumes selling at different average prices, which are the **two most volatile variables to Break Even,** as opposed to **Costs which are secondary.** The accounting gurus have concentrated on costs by theorizing which costs are fixed or variable based upon sales dollars; **this is impossible to accurately formulate because the same cost in many cases have a split relationship, while other costs are in no way related to dollars.**

The fact that Break Even can be modified each week as to volume & price makes it the perfect control for any business. Every business has good & bad weeks, but as long as a week surpasses Break Even, the growth of a business will not be compromised in that particular week. The web site **www.performancecash.com**   includes a **Special Break Even Club.**

Excel Included in this Presentation

The Truth to Share Multiples Based Upon Unit Sales & Unit Prices

See Epilogue

Re-forecasting Operating Earnings from Performance Cash Based

Upon Units

Part of XL100 Series

An Open Letter to Shareholders and Financial Analysts

I need your help; I have been at this by myself for several years and have not cracked the acceptance barrier. Performance Cash is nothing more than what a 16th Century Printer knew about his business before the first CPA was born. The data is readily available every week; it is a matter of Stockholders demanding this information from the Management of their Corporate Investment.

To put hours paid in a simple but powerful and absolute light. If your company paid you **$40.00 an hour (Benchmark# 2)** 2) they are probably selling yours and all other employee hours of the business at an average selling price in the $120 range **(Benchmark #1),** 3) have paid out another **$30.00 in variable material cost,** and another $30 **for all other expenses** to **net $20.00 in Performance Cash (Benchmark # 3).**

Major Corporation do budget, say the above was below sales expectations which were **$125.00** with **$35.00 of labor cost** and **$25.00 in other expenses** for **$34.00 of Performance Cash**. The items are all $5.00 different (except for material) from actual and equal a $14.00 shortfall in total cash, significant but easily made up if one more hour was actually sold as follows.

	Expectations	Actual	Variable
Sales per Hour	$125	$120	$125
Material Costs as %	31	30	31
All Labor Costs	35	40	
All Other	25	30	
Performance Cash (net)	$34	$20	$94

Although column three is only one more hour it is really double the sales volume. Performance Cash of $94, more than offsets the $14 loss in cash because it does represent twice the anticipated volume; one verse two hours. Also, if the original hour were to be sold at $14.00 more or $134, the **entire Performance Cash short fall** would have been offset by price.

My point is rather extreme in the doubling of the sales volume, but the **volume** of units and their **selling prices** are the real driving force of any business. The sale of each and every unit as a composite drives fluctuations in any business from day to day as well as compared to expectations. Meaning, it is not the control of costs that primarily drive profits. Most Accountants serve the cost control function quite well, and are appropriately labeled bean **counters when only performing that function.** What drives profits are: how many units (Earned Hours© transforms units to hours) are sold and at what price; Accountants do not serve this function adequately. That is what Performance Cash accentuates with exact deviations as to: a) Price b) the Mix of Each Type of Product as a Percent of the Whole c) the Volume of Units. Not only does Performance Cash do it for units at one operating facility such as a plant or a retail store, but **PC** does it within each Segment, Division or Product Line of a Major Corporation.

	Expectations	**Actual**	**Variation**	
Example Product A	10 @ $5.00	10 @ $4.00	$1.00 Price	($10.00) bad
Example Product B	8 @ $3.00	12 @ $3.00	4 Units	$12.00 good
Example Product C	5 @ $1.00	5 @ 1.00	none	
Total Sales	23 $79.00	27 $81.00	Dollars	$2.00 good
In Total @	$3.43	$3.00	Mix	($.43) bad

The Financial World, especially the Stock Market, and in particular Analysts deal in Dollars only. **Corporations themselves have no way to exactly pinpoint Cash Flow variations in total dollars let alone EPS in pennies from a sales variance verse expectations in dollars**. Corporate budget specialists of Industry use assumption analysis on their operating statements a couple of weeks after each month end at plant and retail levels. Two of the three (price, volume) variations above are generally known to some extent, but not for each and every product, while the third, mix is assumed. ***Assuming,*** is also done at the Corporate level where the variation becomes more acute and significant as Major Corporations cross into many different Market Segments with great variations in profit margins. Earned Hours© transforms all units for each and every product to absolute and universal hours, with the exact dollar variance to the three elements of sales as well as Cash Flow and Operating Earnings.

You need not understand all of the above but any data is better than what the Investment Community is getting today. In addition, **Earned Hours© verse Paid Hours will** enhance Management's control over its Corporation to allow the rest of the world to sing to the same hymn book. Start harassing your Corporate big wigs for better information. You pick up the newspaper every day and find out how your favorite sport team is doing, there is no reason it can't be done for each and every investment you own.

Fast Food and Restaurants

Performance Cash Specially Defined for the Industry with a Test Case Supported & Useable by XL 18, 19 & 20. The three separate XL worksheets have specific instructions to put them to work next week.

XL 18 of the Appendix contains one printed sheet from a prototype program with a test case that will receive all new data with only a minimum of original entries. It is called the set-up sheet. There is a another sheet which I refer to as the work-horse that will receive data from lines 29 to 47 of set-up, and feed back the results in the lower

portion of set-up, lines 50 to 53. Work-Horse never receives original entries.

Performance Cash Specially Defined for Fact Food and Restaurants

Cash Sales + Credit Sales of the Cash Register less Gross Payroll and all Expenditures from the Cash register. I will refer to it as Net CR PC

There are two principal Drivers for **Net CR PC** whether Positive or Negative. If the two are under control it means that the amount of PC makes sense. There will be times when not much can be done about the results, but I would not expect that to happen until a long way into the use of this specially designed program for the Food and Beverage Industry.

The two principal Drivers are Sales and Labor Costs

Before I get to the meat of the definition let me discount Ingredient Cost that are generally the most significant and consistent cost as a percent of sales. Such costs will only be itemized to the extent that they go through a Cash Register. These costs occupy **line 25** of **XL 18. Line 24** of XL 18 are other cash register payouts. Both expenditures are best controlled in total by the normal weekly Performance Cash Report that you can leave aside until you get sales and labor costs under scrutiny and control.

Labor Cost for all personnel are the most volatile of any business because the effectiveness of the amount paid for **Net CR PC** will always depend upon the number of units sold. The hours of labor and weekly PC are used to establish a base for 100% of any unit forecast for any business. When unit sales deviate from that anticipated the impact on PC is always known in dollars. Whether management realizes it or not, most every business anticipates a level of unit sales volume by how many people are hired before the need. This book has developed programs around the age old 100% (units) concept which most people understand but accountants are reluctant to apply to monthly financial statements, let alone for every week of a business year as well as for every day of a business week. Daily Labor Costs are something less than 100% variable, but the percent will be always be known because even the slowest day of the week has a 100% labor milestone. To know and control the cost of **direct labor** relevant to unit sales has been successfully practiced in manufacturing circles for many many years. **Net CR PC** subjects **all labor** costs for any erratic retail business to the same control as Manufacturing. With Marketing predictability tools that you may currently have under your belt, similar effectiveness can be achieved. Every retail business should know if it had a good or bad day through fact; not feeling or fiction and that goes for the slowest day of the week as well as the best.

While **Net CR PC** only tells when Labor Costs need adjustive action, net Sales dollars are another case, any fluctuation in the trends of price or

85

the volume of units falls directly to the bottom line of PC. The good news you will always know the impact of the average price of each principal product and the unit volume of such each principal product on PC from day to day or week to week.

The Mechanics

By dividing labor hours by units so that each principal product (Dinners, Lunches etc.) sold has a predetermined hourly rate **column G line 13 to 18** and a selling price per hour **(f21)** a specific program has been developed from Monday through Sunday for Restaurants, Fast Food (burgers, sodas etc.) and almost any other sporadic retail business.

For each day of a week and from your input to **lines 5 and 6,** the **Program** sets-up how much you expect to pay per hour as an average to all your employees, $13.60 **(XL 18 e8)** for each week. The program also sets the Performance Cash expected to go through the Cash Register $19,600 **(d26)** in a week. As each day ends when your Cash Register tape indicates $3,000 in sales and you send credit slips for $1000 (I.E) to your Merchant Bank and deposit $1500 (I.E) in your own bank account. You will have a dollar accounting for any difference by the unit price and volume of each principal product and the overall unit sales efficiency of all the personnel paid that day.

From the Appendix set-up page---please observe column B which is Monday's performance.

Sales of $2,500 **(b42)** were considerably below the average for each day of the week $5,143 ($36,000 **(d21)** /7) but the sales per hour of $36.29 **(b43** Benchmark # 1) were actually better than the $36.00 anticipated in the upper grid. Benchmark # 1 **(line 43)** measures the sales price of each principal product sold and the mix of actual results among all the units sold (16,000 in cell **b19**). The dollar improvement $.29 for each hour sold verse $36.00, had little impact as reflected by positive $41 in **b50.** Volume is an entirely different story; all the units sold represent 68.9 **(b41)** hours verse 142.9 **(j41** 1000/7), a 74 hour shortfall for each hour at $36.29 equals $2,684 **(b51)** less Performance Cash for the day. The **slowest day** did not meet the 100% requirements of the **week. Management would know what is require of that day, $13.60** (b41) **x 68.9 or $937 of payroll dollars to be 100% efficient.**

To summarize the above, both **b50 and b51** are out of pockets variance to your cash deposits for Monday. For the balance of the week, Tuesday went unfavorable price wise $35.25 indicated by **c43 being** below $36.00 as reflected by the unfavorable dollar variance of ($106) in cell **c50.** Each day fluctuated and maxed out at $37.87 **(d43)** on Wednesday and finished favorable at $36.40 for the week **(k 41).**

Labor

Such Costs are obligated by a schedule (**columns K & L top**) before you know the extent of business sales in any one week. You can call such costs "out of pocket from an empty pocket" until sales exceed a labor break even. The Case numbers are based only upon personal observations, but an excellent example of the 100% application for a fast food retail outlet or a restaurant. In manufacturing Plant Management knows what people handle the product and they call them direct laborers.

With food serving, many employees are not going near the product but are serving the customer in a variety of ways and must also be considered direct. Bus people, cashiers, hosts, janitors and managers are all on call for serving customers; they certainly make their present known in mass at the large Orlando restaurants by their constant movement, but are they efficient? It is no fault of their own when working with a shortage of customers; but that of management. There must be a link between the scheduling of all personnel, the food available to be served each day, and customer demand in units.

I realize there is a direct sales link to the waiters and waitresses and tips are the best variable cost control of any Industry except maybe for Real Estate Agents and their commissions. All those support people on the restaurant floor and in the kitchen are only partially variable, but they will dictate the profits of the establishment.

The Set Up is all that is necessary to put the Earned Hour Process in Motion

XL 19 starts with a set up page that puts to paper what you already know about the operation. First, are **lines 4 to 8,** how much you paid to your salary and hourly people the previous week and how such hours were disbursed over the week **columns K &L.** Salary people are always recorded at 8 per day, five days a week and 2000 for a year, but they should have flexibility on slow days to perform hourly functions.

Sales

With the sales data on the upper grid of **XL 18,** I realize that this information may not even be available; a best guess is all that is necessary for set-up because before long the truth will be known.

XL has six products listed; they were chosen for price variations as well as what I believe should have different ingredient mark ups. If you know these particulars about the six principal products of your business, unit profits will fall into place. Other than the major six, are not neglected, they are part of the overall mix as the sales per hour of each employee is represented by Benchmark #1, $36.00 per hour **f21.**While I am on the different products and after you have **XL 18** under your belt refer to **XL 14.** Other cuts of the Market can be added to the same Sales dollars with the same customers to be served, by breaking sales reporting down into markets such as (A) children women and men or, (B) morning, noon and night. These are not alternatives, but duplicates that can be called Bx #1A & #1B. Of course once set-up, they would require similar recording on an actual basis in addition to that above.

For restaurants a starting set of products could be: Breakfast, Lunch, Dinners, Buffets, Alcoholic Drinks and Non Alcoholic Drinks. If you don't serve breakfast, you can insert a specialty item such as desserts or appetizers not included in the Dinner price. I consider Buffets a separate and necessary class of product because it must have different profit parameters.

Labor

As a Manager, you probably have a good feel for the requirements of an entire contingent of personnel in the kitchen and on the restaurant floor, but how you align their assignments by day with the support people is the key to profit refinement. As you may already do, call upon key help to do the support functions on slow days to give them their 40 hours per week

You must determine the right 100% balance between paid dollars **line 44** and sold hours **(lines 35 to 41)** that are converted from unit's sales each day. The numbers come right from the set-up grid. **Column K and L line 13 compared to line 41.** Summarized below are the percent of efficiency for each day of the week, with 100% your goal.

| | scheduled hours | | | | |
	salary	hourly	hrs to pay	earned	efficiency
Monday	20	60	80	68.9	86%
Tuesday	20	60	80	93.6	117%
Wed	20	80	100	110.9	111%
Thur	30	100	130	146.9	113%
Friday	30	100	130	182.1	140%
Sat	40	200	240	211.0	88%
Sunday	<u>40</u>	<u>200</u>	240	171.2	71%
	200	800	1000.0	985	98%

I know it is unrealistic to expect prediction perfection regarding sales units, but a pattern will develop by hours earned **(line 41)** for each day of the week, but you can also help direct the pattern by "give always", special prices etc. That is what management is all about, try to match the set-up to the pattern. There is nothing to say hours can't be taken away from those less then 100% days, after all if you can be at 100% on certain days, why not try to achieve the same on all days.

A reminder, only Performance Cash as generated by a Cash Register with Credit Card sales were addressed above. All the detail of Net PR PC is easily summarized on the Deposit Worksheet #4 of the Program. Such a re-cap can also be used on other weekly reports of normal Performance Cash as defined throughout other parts of this book. That definition is worth repeating **Cash Sales + Credit Sales less Cash Out for all Expenditures except Capital Investment & Debt Service**

A Restaurant XL 19

Using the Fast Food Prototype Program I modified the sequence to establish a schedule for the given workforce of **lines 5 through 8** and added a more representative server rate without tips.

Keep a record of unit sales by class and by day with the total dollars for the week and strike an average price for the week for each class of units. Plug d20 to agree with dollars sales for the week to arrive at, **f21 (i.e. $28.65).**

1) Insert units by day on **lines 29 to 34** to agree with units for the week, **column I.**

2) From **line 41** multiply the number of earned hours by **f21** to arrive at sales for the day and plug the last day to agree **with d21** (i.e. $55,000).

3) Again multiply the earned hours of **line 41** by the average cost per hour of labor (i.e. $9.38) to arrive at the daily payroll. Insert the amounts of **d7 hours** (i.e. 1920 in cell **i45** and **f7** dollars (i.e. $18,000) in cell **j45.**

4) Go up to the schedule of hours and fill in current schedule of salary people so that k19 agrees with d4.

5) Fill in the schedule for the hourly personnel to agree with **line 41** when added to the salary.

Results----e 55 agrees with e56 (i.e. $9.38) your program is correct and ready for immediate use to see how efficiency can be improved on slow and best days.

A Specialty Retailer XL 20 (the example assumes no spending variance in labor and other spending for the complete week)

This case was set-up with the **retail price** of each of the principal products in **column c lines 13 to 16.** The example for actual data of **lines 29 to 34** contains unit shortfalls through Saturday for the blouses and pants, the two principal products. I intentionally multiplied **line 41** through Saturday by **Benchmark #1 37.88** to arrive at the sales amounts on **line 42.** Through Saturday, line 50 shows little variance for the week. But check out the unfavorable unit shortfall in dollars (not shown unless added across, $3905) on **line 51** through Saturday.

Sunday had various discounts on a portion of the units for each of the six products as indicated by **h50** on better than average volume for the two major products. The actual results as indicated from **h50 and h51** took price unfavorable for the week $1067 **(i50)** with volume down only $1020 **(i51).**

Conclusion

By setting up the average retail price of the inventory on hand in the upper grid or what a Manager would want to realize from each major product, he or she will know: 1) the actual volume of units being sold 2) the efficiency of the sales force 3) the exact dollar impact on sales due to discounts from the ticket price each day and year to date.

Summary All Retailers

At all times and daily a retailer will always know if the or she had a good day through the efficiency of all the personnel to sell a certain amount of units on each particular day and the dollar impact of the exact selling price of each principal product.

Appendix

The Power of the Hour

The Hour has been an accounting **source of Power** used to standardize the cost of manufacturing individual products for many, many years. The hour brought **consistency** to a variety of cost elements. Even though a plant may only fabricate one part, daily costs (scrap rates etc.) tend to be very inconsistent. I have merely taken prearranged hours of manufacturing personnel & expanded them to include all personnel through administration, distribution & sales of the product & allocated them over sales dollars by price categories or classifications.

I recommend no more than six product differentials or price classes no matter how extensive the Product-Line. You would be surprised how many facilities can be monitored as a **"single-unit"** business. One principal product or dominant product (**Driver)** magnifies the law of large numbers whether a Supermarket or Drug Store (**customers**) a Bank (**depositors**) a **ticket holder** (Theme Park with food & cloths) or **generic** (lbs. square feet or gallons).

The product or service may be contractual (Architect), a retainer (Law Firm), or a random service (Doctor or Maintenance Company) or one facility that houses two (hotel with dining) distinct business segments or that of a Publisher (copy sales & advertising). A non-volatile monthly driver such as a membership (golf) using many facilities (Country Club) including its dining room may also drive an entire business.

My first book *How to Predict Year-End Cash* designed averages from the two most **significant quantities** of a business, **HOURS PAID & UNITS (or Drivers) SOLD.** Gathered on a year to data basis they behave like all large numbers; **hard to change once the average & its resulting benchmarks get momentum.**

There are many accounts that comprise what I use to call a P & L statement. Tradition has been geared to regulatory requirements that have obscured numbers that are capable of laying a business bare. Sales **(unit-hours)** connected to **(dollar-totals)** the key operating accounts of a P&L and/or Cash Flow Statement can monitor changes to the principal balance sheet resource, **Cash.** Therefore a business can be tracked by the **employee hour** (first planned & then paid) & expose major profit & loss changes.

Notes For Author

Break Even, XL #3

Before Performance Cash came along, there was an accounting theory that has **never been accurately** put to practice----**Break Even. Any forecast of PC is convertible to a Weekly Break Even point**. The Break Even of PC is primarily based upon different unit volumes selling at different average prices, which are the **two most volatile variables to Break Even,** as opposed to **Costs which are secondary.** The accounting gurus have concentrated on costs by theorizing which costs are fixed or variable based upon **sales dollars; this is impossible to accurately formulate because the same cost in many cases have a split relationship, while other costs are in no way related to dollars.**

The fact that Break Even can be modified each week as to volume & price makes it the perfect control for any business even when exceeding break even annually. Every business has good & bad weeks, but as long as a week surpasses Break Even, the growth of a business will not be compromised in that particular week.

In addition, the **Year to Date excess of Break Even** will be at a **rate per hour** based upon units sold & hours **paid** to **all your employees. A quality performing CEO will be proud to report both performance indicators to his shareholders.**

The web site **www.performancecash.com** includes a **Special Break Even Club.**

When there are deviations **(always)**--- to a plan, budget or a revised estimate, the **exact variances** are documented by **unit volume,** (how many), **unit price** (by major product), & **third, cash spending** in four major categories. They are 1) labor, 2) fringes on labor, 3) all other operating expenses & 4) a very large variable as a lump sum & percent of sales: all material, raw material, ingredients & subcontracting costs. The expenses include **no accounting journal entries**; that is what true cash flow is all about. There is nothing else to know about your business, & **it is every week with no Accountant needed to interpret the results.**

The Most Volatile Variances

The most important weekly variables about any business are:

The Unit Price Variance

Actual Benchmark #1 **(e33)** $26.10 per Earned Hour verse

    that anticipated $26.60 **(f17)** x the number of Hours Earned 1123

    (f31) Equals--- ($566)--- unfavorable (h32)

The Unit Volume Variance

Actual Hours Earned 1123 **(f31)**

    minus the hour monitor 1000 **(d3&f15)** x Benchmark #1 $26.60

    (f17) Equals--- $3,264---favorable (h33)

The All Payroll/Labor Efficiency Variance

Actual Benchmark #2 **(e34)** $11.67 per Hour Paid **(/1200 c34)** verse

    that anticipated $13.60 **(e3)**

      Equals--- 86% (e35) efficient---- $11.67/$13.60

Performance Cash or Cash Flow with Credit Sales Variance

Actual Benchmark #3 **(e38)** $1.78 per Hour Earned verse

    that anticipated $0.00 break even **(d22)**

      Equals---growth --- $1.78/$0.00, also a percent when applicable

To be certain that **you** capture & use the back bone of the entire book, two methods to the Prototype graphic follow. The first is to duplicate my exact numbers with all the formulas on your own excel spreadsheet. The second is to create your own Program with your numbers. That approach is in sheet or slide sequence & follows. Both were separately tested by non-accountants that did not use Excel for their work. If approach two is used the spending formulas of column H must be added.

In summary---A) the Truth & Performance Cash results of the lower grid reflect a comparison of the same cash flow ($2,000) for each unit sold (e38/f31) verse each hour paid (e39/c34).---B) PC results give dollar to dollar spending variances (Column H) by major categories compared to any anticipated results.

From either Prototype, open and save a "Sand Box Break Even ". Although my upper grid is a break even, it should not be when you do your grid. Put your best guess or a previous week of actual in the cells of your upper grid, then try to make the bottom grid break even. The upper grid can remain permanent for the balance of the year, it is the **Truth** of the bottom grid that you are after.

You can test your file of the program by changing the hours of **d3** to anything you wish. Neither the truth of e39 or column H changed in relation to each other (in particular the two sales variances h32 & h33). The hour set up of the upper grid is merely the measuring monitor to get to the **Truth for each Hour Paid & each Unit Sold.**

Return **d3 to its original amount & change** the payroll of **f3.** The top grid & d22 changed to reflect more or less labor costs. Only the labor variance of **h34** changed.

Change the hours paid **(c34)** for the week to less than **f31.** Bx # 2 will be 100% plus or minus an efficiency factor **(e35)** as reflected by the dollar value of **e34;** because the **production output (lower grid column B verse upper B)** for the week met or exceeded expectations.

The Last & Most Important Observation

Performance Cash, cell **d38** divided by **Hours Earned, f31** = the rate from **Units actually sold.**

Cell d39 (the same amount as **e38**) **divided** by the **Hours** of **c34** is the Performance Cash Return for Each Hour **Paid** to all employees. **A more self evident Business Truth is not available** but obviously different from the Earned Hour Rate. Go back & change **d3 so that** both **e38 & e39** are equal as a rate per hour. The relationship changed because the hour monitor of set-up changed, **not sales or the cost** within the upper grid.

Management would communicate the actual **Year to Date PC per Paid Hour** in comparison to any prior period for the benefit of any interested party. Depending on its rapport with Shareholders & other Economic conditions, Management could bring **e38** into its comments along with its ability to improve the situation as outlined from line 32 to 36 column H.

Again play with **c34** & **d34** up or down for **e39** results. If the amount per hour of PC is more than **e38,** Management did well for the units actually sold during the week & had a **better return for each hour paid.** If less than, Performance Cash & Management could have done better for the units that were sold.

	A	B	C	D	E	F	G	H	I
1	Power of the Hour		*understanding will derive so much from so little*						XL #3
2		*Prototype*		hours	hourly	no fringes			
3	All Labor		Bx # 2	1,000	$13.60	$13,600			
4	is connected to the upper grid by— d3 to f15— & —— f3 to d18								
5	*my data now, your's later*								
6	this upper or set-up grid is your best guess at Column B x Column C								
7		weekly	price per	weeks	sales	% hour	to be		
8		units	unit	sales	allocation	allocation	earned		
9	burgers	1,400	$ 4.00	$ 5,600	22%	225	0.1604	col F / col B	
10	sandwich	1,000	$ 5.00	$ 5,000	20%	200	0.2005		
11	platters	600	$ 7.00	$ 4,200	17%	168	0.2807		
12	salads	400	$ 3.00	$ 1,200	5%	48	0.1203		
13	coffee	4,000	$ 1.00	$ 4,000	16%	160	0.0401		
14	soda/beer	1,300	$ 3.80	$ 4,940	20%	198	0.1524		
15		8,700		$ 24,940	100%	1,000			
16	other sales (part of the mix)			$ 1,662	*this line can be used as a plug*				
17		totals		$ 26,602		$ 26.60	Bx #1	d17 / f15	
18	spending labor			$13,600					
19	spending labor fringes d18 x %			$3,400					
20	spending all other expenses			$5,000					
21	spending inventory purchases d17 @ %			$4,602					
22		Performance Cash		$ (0)		$ (0.00)	Bx #3	d22 / f15	
23									
24	Any Week's Actual		shaded is new data			col b X e. hours earned			
25	burgers	1,500	$ 4.20	$ 6,300	0.1604	241		d33 / f31	
26	sandwich	1,000	$ 5.00	$ 5,000	0.2005	200			
27	platters	600	$ 7.00	$ 4,200	0.2807	168			
28	salads	400	$ 3.00	$ 1,200	0.1203	48		d34 / c34	
29	coffee	4,000	$ 1.00	$ 4,000	0.0401	160			d38 / f31
30	soda/beer	2,000	$ 3.50	$ 7,000	0.1524	305	difference of grids		
31		9,500		$ 27,700		1123	Perf. Cash	$ 2,000	d39 / c34
32	other sales			$ 1,600			price	$ (566)	
33		totals		$ 29,300	$ 26.10	Bx #1	volume	$ 3,264	
34	paid hours & dollars =Bx #2	1200		$14,000	$11.67	Bx #2	labor	($400)	d18 –d34
35	spending, labor fringes			$3,500	86%		fringes	($100)	d19 – d35
36	spending,all other expenses			$4,800			expenses	$200	d20 – d36
37	spending, inventory purchases			$5,000			inventory	($398)	d21 – d37
38		Performance Cash		$ 2,000	$ 1.78	Bx #3		$ 2,000	
39	Bx #3 as stock multiple-- d39 / c34			$ 2,000	$ 1.67	Bx #3	*the truth of any business*		
40		price		(e33 ($26.10) minus " f 17($26.60) " f 31(1123)= ($566)					
41									
42				actual Bx #1		estimate Bx #1		actual volume	
43		volume		f 31(1123) minus f 15(1000) " f 17($26.60) = $3,264					
44									
45			actual volume		estimated volume		estimate Bx #1		
46									

Monitor Set-Up
Sheet #1

1-- Set-Up---Insert in d3 a weeks labor hours (salary @40) & in f3 all labor cost.
2--Allocation---Need Product, Unit, & Price detail of columns A, B & C to arrive at column D---each Product is taken as a percent of the whole e15. Each Product's percent is multiplied by d3 to arrive at Column F. You need not list 6 Products but I would not exceed 6 at the outset.

Create file on one sheet with your input to exactly the same cells					
Break Even Prototype		hours	hourly	no fringes	
All Labor		1,200	$13.33	$ 16,000	
		d3		f3	
				1200	1*d3
spending labor		$16,000	1*f3		

3--Hours Assigned---Each Product's hours of column F are divided by units to arrive at the **year's** Earned Hour© rate when a unit is sold, column G.
4--Paste proof---Take the yearly rates of each product and line up with the appropriate Product on line 25 column E. Copy exact numbers for each product from the upper cell Columns B & C on lines 25 to 30. Make sure f31 agrees with g31.

5--Benchmarks Upper Grid---insert an estimate of other "sales" in d16, Benchmarks will automatically calculate, Bx #1 f17 Bx #3 f22.
5--Benchmarks Lower Grid---Actual data for a complete week from sheet # 7 (not shown but is part of Appendix Program). Automatically calculates, Bx #1 e33 Bx #3 e38.
5--Labor Benchmark Lower Grid---Insert Actual Labor Hours c34 & Costs d34 from payroll report. Automatically calculates, Bx #2 e34 & e35 Bx #3 e39.

6--Upper Grid Cash Outlays d19, d20, d21 insert best guess
6--Lower Grid enter from check book d35, d36, d37

Product Set Up
Sheet #2

	A	B	C	D	E	F	G
1	Power of the Hour						
2	Break Even Prototype			hours	hourly	no fringes	
3	All Labor			1,200	$13.33	$ 16,000	
4							
5							
6	only change shaded cells						
7		weekly	price per	weeks	sales	% hour	
8		units	unit	sales	allocation	allocation	
9	burgers	1,400	$ 4.00	$ 5,600	22%	269	d3*e9
10	sandwich	1,000	$ 5.00	$ 5,000	20%	241	
11	platters	600	$ 7.00	$ 4,200	17%	202	
12	salads	400	$ 3.00	$ 1,200	5%	58	
13	coffee	4,000	$ 1.00	$ 4,000	16%	192	
14	soda/beer	1,300	$ 3.80	$ 4,940	20%	238	d3*e14
15				$ 24,940	100%	1200	
16							
17							
18	spending labor			$16,000			

Earned Hours Assigned
Sheet # 3

	A	B	C	D	E	F	G
1	Power of the Hour						
2	Break Even Prototype			hours	hourly	no fringes	
3	All Labor			1,200	$13.33	$ 16,000	
4							
5	no changes necesssary						the key
6						column F/column B	
7		weekly	price per	weeks	sales	% hour	to be
8		units	unit	sales	allocation	allocation	earned
9	burgers	1,400	$ 4.00	$ 5,600	22%	269	0.1925
10	sandwich	1,000	$ 5.00	$ 5,000	20%	241	0.2406
11	platters	600	$ 7.00	$ 4,200	17%	202	0.3368
12	salads	400	$ 3.00	$ 1,200	5%	58	0.1443
13	coffee	4,000	$ 1.00	$ 4,000	16%	192	0.0481
14	soda/beer	1,300	$ 3.80	$ 4,940	20%	238	0.1828
15				$ 24,940	100%	1,200	
16							
17							
18	spending labor			$16,000			

The Proof
Sheet # 4

	A	B	C	D	E	F	G
7		weekly	price per	weeks	sales	% hour	to be
8		units	unit	sales	allocation	allocation	earned
9	burgers	1,400	$ 4.00	$ 5,600	22%	269	0.1925
10	sandwich	1,000	$ 5.00	$ 5,000	20%	241	0.2406
11	platters	600	$ 7.00	$ 4,200	17%	202	0.3368
12	salads	400	$ 3.00	$ 1,200	5%	58	0.1443
13	coffee	4,000	$ 1.00	$ 4,000	16%	192	0.0481
14	soda/beer	1,300	$ 3.80	$ 4,940	20%	238	0.1828
15				$ 24,940	100%	1,200	
16							
17							
18	spending labor			$16,000			
19						paste	
20							
21							
22	only change shaded cells						
23							
24	Proof	re-enter values from B & C above				col b X e. hours earned	
25	burgers	1,400	$ 4.00	$ 5,600	0.1925	269	
26	sandwich	1,000	$ 5.00	$ 5,000	0.2406	241	
27	platters	600	$ 7.00	$ 4,200	0.3368	202	
28	salads	400	$ 3.00	$ 1,200	0.1443	58	must agree
29	coffee	4,000	$ 1.00	$ 4,000	0.0481	192	with f31
30	soda/beer	1,300	$ 3.80	$ 4,940	0.1828	238	set-up
31				$ 24,940		1200	1,200

The Benchmarks
Sheet # 5

	A	B	C	D	E	F	G
5	only change shaded cells						
6							
7		weekly	price per	weeks	sales	% hour	to be
8		units	unit	sales	allocation	allocation	earned
9	burgers	1,400	$ 4.00	$ 5,600	22%	269	0.1925
10	sandwich	1,000	$ 5.00	$ 5,000	20%	241	0.2406
11	platters	600	$ 7.00	$ 4,200	17%	202	0.3368
12	salads	400	$ 3.00	$ 1,200	5%	58	0.1443
13	coffee	4,000	$ 1.00	$ 4,000	16%	192	0.0481
14	soda/beer	1,300	$ 3.80	$ 4,940	20%	238	0.1828
15				$ 24,940	100%	1,200	
16	other sales (part of the mix)			$ 1,662			
17				$ 26,602		$ 22.17	Bx #1
18	spending labor			$16,000			
19	spending labor fringes d18 x %						
20	spending all other expenses						
21	spending inventory purchases d17 @ %						
22		Performance Cash		$ 10,602		$ 8.84	Bx #3
23							
24	Actual	data from daily sheet 7			col b X e, hours earned		
25	burgers	600	$ 4.00	$ 2,400	0.1925	115	
26	sandwich	600	$ 5.00	$ 3,000	0.2406	144	
27	platters	600	$ 7.00	$ 4,200	0.3368	202	
28	salads	600	$ 3.00	$ 1,800	0.1443	87	
29	coffee	600	$ 1.00	$ 600	0.0481	29	
30	soda/beer	600	$ 1.50	$ 900	0.1828	110	
31				$ 12,900		687	
32	other sales		sheet 7	$ 1,400			
33			totals	$ 14,300	$ 20.81	Bx #1	
34	paid hours & dollars		1200	$ 14,000	$11.67	Bx #2	
35	spending, labor fringes				88%		
36	spending,all other expenses						
37	spending, inventory purchases						
38		Performance Cash		$ 300	$ 0.44	Bx #3	
39		Performance Cash		$ 300	$ 0.25	Bx #3	

109

The Variances
Sheet #6

	A	B	C	D	E	F	G	H
5	only change shaded cells							
6								
7		weekly	price per	weeks	sales	% hour	to be	
8		units	unit	sales	allocation	allocation	earned	
9	burgers	1,400	$ 4.00	$ 5,600	22%	269	0.1925	
10	sandwich	1,000	$ 5.00	$ 5,000	20%	241	0.2406	
11	platters	600	$ 7.00	$ 4,200	17%	202	0.3368	
12	salads	400	$ 3.00	$ 1,200	5%	58	0.1443	
13	coffee	4,000	$ 1.00	$ 4,000	16%	192	0.0481	
14	soda/beer	1,300	$ 3.80	$ 4,940	20%	238	0.1828	
15				$ 24,940	100%	1,200		
16	other sales (part of the mix)			$ 1,662				
17				$ 26,602		$ 22.17	Bx #1	
18	spending labor			$16,000				
19	spending labor fringes d18 x %							
20	spending all other expenses							
21	spending inventory purch d17 @ %							
22		Performance Cash		$ 10,602		$ 8.84		
23								
24	Actual	data from daily sheet 7				col b X e, hours earned		
25	burgers	600	$ 4.00	$ 2,400	0.1925	115		
26	sandwich	600	$ 5.00	$ 3,000	0.2406	144		
27	platters	600	$ 7.00	$ 4,200	0.3368	202		
28	salads	600	$ 3.00	$ 1,800	0.1443	87		
29	coffee	600	$ 1.00	$ 600	0.0481	29		
30	soda/beer	600	$ 1.50	$ 900	0.1828	110	difference of grids	
31				$ 12,900		687	Perf. Cash	$(10,302)
32	other sales		sheet 7	$ 1,400			price	$ (932)
33		totals		$ 14,300	$ 20.81	Bx #1	volume	$(11,370)
34	paid hours & dollars		1200	$ 14,000	$11.67	Bx #2	labor	$2,000
35	spending, labor fringes			0	88%		fringes	$0
36	spending,all other expenses			0			expenses	$0
37	spending, inventory purchases			0			inventory	$0
38		Performance Cash		$ 300	$ 0.44	Bx #3		$(10,302)
39		Performance Cash		$ 300	$ 0.25	Bx #3		

Dollars as Units, XL 4

Just like **Units** are convertible to hours, **Dollars** can also be converted to hours. There are several precautions to avoid so that each class of sales dollars has a different earned hour rate. Each class of sales dollars or Product Segment (i.e. Manufacturing, Retail, & Software Services) from an Annual Report is entered on the standard upper **sales connection grid** along with the labor hours for that particular Segment. Hours of Central functions such as a Corporate Office or a Research Facility that have no operating sales would be accumulated & allocated to each Segment. Then the total hours of the entire Corporation are allocated to each Segment by sales dollars. All the payrolls of the Segments & the Corporate Office plus Research, if that were the case, would be lumped together for one Benchmark #2 rate to measure the unit/sales efficiency of the entire Corporation.

In addition the Benchmark #3 Rate per Hour will measure the efficiency of the entire Corporation to generate Operating Cash Flow.

Another significant wrinkle that should be added to the upper connection grid is the estimated sales increases or decreases as a percent compared to the previous year. Each Segment could also have its own unit or set of units to drive its own Performance Cash.

Martin D'Amico

Modeling for Start-Ups, XL 5

The uniqueness of Earned Hours© for Modeling lies within Benchmark #1….. **Sales per Hour in the last year or the so called finish line of a Plan…..** Even though the Sales per Hour Rate is based upon the **payroll hours at the finish line,** Bx #1 retains its reliability & integrity as a Benchmark for all years of the Plan **as to the price of product**--------At the same time Benchmark #2 (Labor per Hour) & Benchmark #3 (Performance Cash per Hour) establish the integrity for payroll & cash efficiency per hour **actually paid** during each year leading up to the last. In effect, once the **initial Model connection** is finalized with the detail of the intermittent years, only the Benchmarks are necessary when summarizing Long Range Plans. Operating Management is measuring at **day one, year one** against the final year of the Model & the Long Range Plan **during each Current Year.** Next year the process starts again.

There is a large & extremely vulnerable group of people who start a business from scratch; the businesses are referred to as, "Start-Ups." A Start-Up is a very difficult undertaking. Generally there are two major obstacles facing people who start such a business. They have not done it before, & probably have had no experience at being in charge of a complete operation. Add that to the uncertainty of the product and/or service & you have a formula for disaster. A tracer or monitor for those operating years is absolutely necessary because I guarantee that operations will only remotely resemble what was originally envisioned for the Start-Up.

XL 5 can also be used for Long Range Planning when your products are well defined by price & volume. If you only have a general idea as to specifics, XL 6 to X8 may be a better approach especially if there are per square foot revenue restrains at the facility.

Martin D'Amico

Long Range Planning the First Time with Performance Cash
The holy grail of Planning, XL 6-8

Some companies don't consider themselves Start-Ups, but are still not generating enough cash to sustain or grow a business. I will give such companies the benefit of the doubt & call them immature; a business can not tolerate an inadequate cash flow very long. Likewise, a purchaser of a mature business needs a fresh approach to an acquisition. Finally all businesses could use a fresh approach; it is called Long-Range Planning.

With this program referred to as the holy grail of planning, I have managed to connect payroll hours to another fundamental & universal (sq meters also compatible) unit of measurement. All businesses occupy square feet & they are generally revenue sensitive whether a facility is leased or owned. **More importantly Square Feet & Hour amounts allow 100% to be equated with any sales & cash flow forecast.**

The link to square feet is ideal for Long Rang Planning. It need not be precise & is a logical first step in planning most Financial Projects. The program is put in motion from a future year of a LR Plan (could also be an estimate or a budget). Square feet @ a selling price per square foot is assumption #1 while labor @ its labor cost per hour assumption #2. The numbers without the money value are probably the two most significant quantities of most businesses. The two significant percents of most businesses round out the four assumptions included in the holy grail. The percents are fringe benefits based on labor costs & raw materials based on sales. Once the assumptions regardless of the eventual accuracy, are inserted in the first few lines, the grail will be put to work. Management will have criteria from day one to record next week's sales & check results that will produce a variance report based upon how many units are sold & at what price. The **exact (the objective of the holy grail)** variances short of that projected will present objective & current data not obscured by accounting theory. This will enable management to address sensitive areas in order to change future results. Exceeding that projected with those variances will give insight to capitalize on what the business is doing right.

Performance Cash for the Entrepreneur, XL 11

This is the Standard Upper Grid Connection with a lower Grid of Actual & its Variance Generator. But, there is an adaptation of the Line Item "other sales actual". It includes a link to deposits that produces a traditional Cash Flow Statement. The program also includes a link to a Check Book and/or Credit card.

Cost Center Control, XL 12

This is the Standard Upper Grid Connection with a lower Grid of Actual & the Variances generated, but without Sales Dollars. It would be used for Major Departments of a Corporation that process a mass of similar transactions. If used for **different classes** of the same transaction, the entire department's hours would need a head count allocation of hours that was not based upon the percent share of the units produced. Do not use an arithmetic unit allocation, that would produce the same earned hour rate for each class of production.

Single/Sole Product Business XL 13

You would be surprised how many Companies can be considered single service or a sole product business. Disney & how many Visitors come through the gate drive the Revenues of the Theme Park & every other Retail Outlet on the grounds. **Cubic Feet would drive a Moving or Freight Company, Daily Depositors for a Bank, Gallons for a Gas Station/Convenient Store, Memberships (different rates, old & new) for subscribers, Rounds of Golf, Tax Payments for the U.S Government, Passengers or Air Miles for an Airline.** The sole product concept can also be used for multi product businesses by making the Unit of the business, "customers" **(supermarkets) or** "transactions" **(banks again).** The most important product of a business may also serve as the sole product (blouses for a specialty retail shop, hamburgers for fast food, bagels for a bagel shop, dinners for a restaurant, computers for a Dell, visitors at a ball park & visitors to a web site). **I.E.** The principal individual product, blouses can drive the sales & mix of all products with a sales rate per labor hour that includes blouse sales plus all other dollar sales divided by the **unit sales of blouses or computers only. This means the actual sales of blouses or computers need be the only one tracked on a daily basis.**

Benchmark #1 of a *sole product* operation is a **composite mix indicator** with the total unit production or sales volume of the principal product controlled by the **Cost per Earned Hour, Benchmark #2.** Bx #2 measures the cost & the efficiency of the entire labor force relative to all the

unit sales of the one product (generic or otherwise) with an hourly rate that can be compared to the average hourly rate actually paid to each & every employee.

Other Cuts of the Market Place XL 14

This is the Standard Upper Grid Connection with a lower Grid of Actual & its Variances. A standard class of products such as burgers, sandwiches, drinks etc, was changed to read Men, Women, & Children. The program does not need the same number of unit classes but should **tie into the same sales dollars of the standard grid.** The classes of products could have a third cut: Morning, Noon & Evening with the number of Benchmarks distinguished by letters such as Benchmark #1A for standard, #1B, & #1C. If the set-up were truly extended to other markets, the actual data would have to be accumulated by each class; easily accomplished with variable coding. It could also be done on a sampling basis periodically. When used for Geographic's--- unit sales by location will be the results.

Standard Margin & Variances To
or
Managing Mix Benchmark #5 (new) XL 15

To understand Mix is to manage it. Some elements of mix are: the mix of ingredients within a single unit, the mix of sales dollars within a business, the mix of sales units within a business, the mix of material margins within a business. Combine all those variables with, **knowing** that mix, **budgeting** that mix, then try to **realize** the variance between what you planned & the actual mix of your business. It is hard to say if you will need a **mixed** drink, before or after, when you find out what can be the cash impact on your company. Because of the many variables I won't attempt to define Mix; let me say that it is subject to **each** customer's demand, then **each** company's response to that demand every hour of the day; that is volatility. More important, **Earned Hours** will give you the means to measure mix as well as volume, that is dynamic management. Benchmark #5 operates much like #2 & #3, but instead of sales being allocated, the dollars of standard material margin are allocated & each unit will earn another rate that indicates the volume of standard material margin & the deviation to a budget. Yes if you insist, Product Margin can be net of Labor & Overhead standards.

A correlation to managing mix can be observed by watching a NFL football game. It starts with a game plan or budget with the premise: *nothing is going to go as planned.* The CEO & his Staff of coaches are in touch by headsets after every play, reacting to what the opposition or the **customer is revealing about that day's performance.** They can't wait for the game films or monthly statements to react; the CEO & Staff must be on top of *change* for the next series of downs or customers.

Auditing Program of Operating Earnings, XL 16

A preliminary audit mechanism starting with a week's payroll review of hours & dollars creates a Volume, Price & Spending Variance Report versus any other quarter. The Report is a combination of CPA Public experience with Controller routines, created as an **operating audit procedure** to support Balance Sheet Certification. The analysis will guide the Audit Manager to scope & materiality of variances that will surpass the client's own knowledge of his or her business. That is, if you the auditor has not introduced the client to Earned Hours.

I began my financial career as an auditor for J.K. Lasser & Co & ran some good size audits (McGraw Hill Book Company) by the time I left five years later. I have a feeling that much has not changed from 40 years ago & still lacking in the scope of operating audits. During the preliminary part of any audit, I always had a very good feel for the client's Operating Earnings. Such awareness is so important for materiality & scope. Without earning awareness, it is almost like being an umpire without knowing the score of the game. Not that it means an auditor would take different steps, but he or she would know where to concentrate. Sampling for a Revenue Statement only proves that a period in time is correct, not an entire year.

Discontinued Operations or Massive Layoffs
plus
Justification of Fixed Assets Acquisitions XL 17

Plans to layoff 5,000 employees & discontinue certain complete operations look great on paper & sometimes get Management off the hook in recovering from a sustained sub-par performance. Of number **one** importance, what other criteria was used for ascertaining whether the move is truly Benchmark prudent? **Two,** after execution, was the layoff worth it financially. Before & after Benchmarks (One, Two & Three), lay out the parameters for all to observe with a defined audit trail. The same approach can be used when the purchase of fixed assets has financial justification attached to its feasibility.

Fast Food/Restaurants/Specialty Retailers (see page 83 of text) XL 18 -20

Performance Cash Specially Defined for Fast Food & Restaurants

Cash Sales + Credit Sales of the Cash Register less Gross Payroll & all Expenditures from the Cash register, referred to as Net CR PC

There are two principal Drivers for **Net CR PC** whether Positive or Negative. When **Sales & Labor Drivers** are under control it means that the amount of Performance Cash makes sense. There will be times when not much can be done about the results, but I would not expect that to happen until a long way into the use of this specially designed program for the Food & Beverage Industry.

The Program scrutinizes Daily Sales & Daily Labor Cost. Ingredient Cost are generally the most significant & should be the most consistent cost as a percent of sales. But, such costs will only be summarized in this program to the extent that they go through a Cash Register. These costs occupy **line 25** of **XL 19.**

In addition line 24 of **XL 19** is for other cash register payouts. Both line item expenditures are best controlled in total by the normal weekly Performance Cash Report that you can leave aside until you get Benchmark #1 & Benchmark #2, Sales & Labor Costs under complete scrutiny & control.

Epilogue

	A	B	C	D	E	F
2	Annual Sales Forecast by the Earned Hour©					
3	Annual Performance Cash by the Paid Hour					
4	Earned & Paid Hours are Equal before a Year Starts					
5		Week's Average		Segment Payroll		
6	Segment	Sales		Weekly	Hours	
7	Man'f	$ 300,000		$ 82,000	8,000	
8						
9	Retail	$ 450,000		$ 98,000	9,000	
10						
11	Software	$ 450,000		$ 60,000	3,000	
12		$ 1,200,000		$ 240,000	20,000	
13	Corporate Office			$ 40,000	1,600	
14	Hours	21,600		$ 280,000	21,600	
15	Sales	$ 55.56		by the Earned Hour©		
16	Cash	$300,000		Performance		
17		$13.89		by the Paid Hour		

All that it takes to get to the Truth

x <u>EPS of SEC TRADITION</u>

<u>**PERFORMANCE CASH</u>

x Subjective Quarterly Numbers

** Raw Weekly Universal Data

x Needs Annualization

**Year-to-Date Moving Average

x Accountable To & For Whom?

**Accountable to a Unit Forecast

x Burdensome to Re-Forecast

**Naturally Annualized

x Understood Only by Accountants

**Explainable by a CEO

x Industry Comparisons Not Always Valid

**Even Absolute for Conglomerates

x No Cause & Effect

**Cause & Effect in Pennies

x Inflationary

**Reflective of Business Cycles

x Monthly Break Even Is a Theory

**Weekly Break Even Is an Equation

x Understood Only by Accountants

**Logical for Analysts

x Is a Return On a Static Investment

**Is a Return On a Dynamic Investment

x Bottom Line Subject to
    Inventory Accounting

**Bottom Line Only Subject to Unit
    Pricing & Volume + Spending

x Matches Cost with Revenue

**Presents Expenses as Incurred

x Understood Only by Accountants

**Relative to a Shareholder's Hourly Wage

x Depreciation, Interest & Tax Sensitive

**No Deprecation, Interest or Income Tax

x No Collaborating Benchmarks
    Except Percent of Sales

**Three Other Benchmarks to Assess
        Officer & Director Performance

X Accounting Driven by Accountants

*****Accounting Driven by Marketing

	D	E	F	G	H	I
44		Hours Paid		280,000		310,000
45	*Operating Earnings*		13	Quarterly		15 Weeks
46	*Verse*		wks	Operating		Perf. Cash
47	*Performance Cash*			Earnings		as Reported
48	*Always in View*			Per Hour		Weekly
49				Paid		per Hour Paid
50		Forecast		$ 13.89		$ 13.89
51	Favorable Unit Pricing			$ 0.49		$ 0.88
52	Unfavorable Unit Volume			$ (3.92)		$ (1.47)
53	Favorable Spending			*$1.33		$ 0.25
54	45% is a representative rate			$ 11.79	YTD	$ 13.55
55	for depreciation interest & tax			$ 5.31	➞	**
56	$5.9 mil annualized / 8 mil OS			0.743		
57	last 2 wks better – i51 & i52			$ 14.86	20X	
58	13.55/11.79=115%x $14.86					$ 17.07
59	multiplier 17.07/13.55					1.26
60	*This variance is still influenced by Inventory & Accrual Accounting					
61	**Shown for reconciliation purposes only, when Performance Cash					
62	is accepted this accounting theory will be taken out of the game.					

Letter #1-- Shareholders Speak Out

To V.P Shareholder Relations, ABC Corporation

Regarding: A Book by Martin D'Amico that exposes 3 Ps of every business

Product, People & Performance Cash.com

Dear Mr. Vice President

I have been observing the Enron mess unravel in both the Legislature and the Courtroom with a noticeable silence from the Accounting Profession. As a shareholder of your Corporation I now want to be heard.

Quarterly EPS is Penny Wise & Equity Foolish. There is no accounting for cause & effect in pennies. **Weekly Performance Cash, an hourly rate in dollars & cents has cause & effect in pennies.**

One Weekly Benchmark that reflects Performance is raw data **before Accountants transform the same amount to their general and subjective standards. Even I understand an hourly moving average that can be instituted and executed with a minimum of effort when Accountants are so directed.**

In spite of the US Copyright Office not opening any mail for months after 9/11, Martin D'Amico finally got his simple Earned Hour© Program Certified. The uniqueness of the Copyright---a certain volume of units (100%) translates into Operating Cash, allowing the same flow of cash to be **simultaneously** expressed as a return of hours paid and a return on each unit

128

sold. In the long run the simplicity of Earned Hours© representing units sold will be a boon to the Accounting Profession as they join with General Management to make unexpected numbers legitimately better.

Two significant segments of the Accounting Profession blessed Earned Hours. Management Accounting Magazine which is published by the Institute of Management Accountants featured the **cash link to hours paid** as a monitor for both daily results and business growth in its January 1999 issue. The IMA approximates 70,000 members; the article met their self-study quiz requirements. National Public Accountant, November issue, reviewed the hour process as included in the author's first book quoting *"he shows how cash, the great variable is connected to the two principal statistics of a business, sales units and labor hours"*.

When that hourly dollar & cent average of your Corporation begins to fall and stays below unit expectation you can no longer claim surprise, either make spending cuts or accurately re-forecast results based on new unit sales. D'Amico also has a program that can reconcile Weekly Performance Cash with the Monthly Operating Earnings of SEC Tradition.

Please reveal your Actual Performance Cash per Hour in comparison to your latest forecast each Monday afternoon via your web-site.

A Shareholder of ABC Corporation

Letter #2---CEOs Beat the Shareholders to the Punch
for Incremental Equity Value

To the Shareholders of ABC Corporation

Regarding: A Book by Martin D'Amico that exposes 3 Ps of every business

Product, People & Performance Cash.com

Dear Shareholder:

Because Quarterly EPS is Penny Wise & Equity Foolish and there is no accounting for cause & effect in pennies, **Weekly Performance Cash, an hourly rate in dollars & cents that has cause & effect in pennies has been instituted at ABC Corporation.**

One Weekly Benchmark that reflects Performance is raw data **before Accountants transform the same amount to their general and subjective standards. Even I, a non-accountant can understand an hourly moving average.**

The uniqueness of Performance Cash---a certain volume of units (100%) translates into Operating Cash, allowing the same flow of cash to be **simultaneously** expressed as a return of hours paid and a return on each unit sold. In the long run the simplicity of Earned Hours© representing units sold will be a boon to the Accounting Profession as they join with General Management to make unexpected numbers legitimately better.

Please visit our Web Site & observe the Year to Date Performance Cash Benchmark as well as that for the last week. When it begins to vary below expectation we will apprise you of the reason why with our plans to correct the situation. When warranted a new Forecast with its Performance Cash Benchmark will be posted for your consideration.

Sincerely,

CEO, ABC Corporation

Letter #3--- Public Accountants Lead Clients to the Truth in Reporting

To the Clients of XYZ Partners

Regarding: A Book by Martin D'Amico that exposes 3 Ps of every business

Product, People & Performance Cash.com

Dear Client

In spite of the Copyright Office Certifying Martin D'Amico's Earned Hours© Program it took a while for the Accounting Profession to believe there could be anything new in the Presentation of Operating numbers for Management's sake. The uniqueness of the Copyright---a certain volume of units (100%) translates into Operating Cash, allowing the same flow of cash to be **simultaneously** expressed as a return of hours paid and a return on each unit sold. We would like to now present the simplicity of Earned Hours© to your Entire Organization where Accountants stand to be the big winners when they join with General Management to make unexpected numbers legitimately better.

Mr. D'Amico expresses the Operating Results with Weekly Benchmarks that reflect Performance as raw data **before Accountants transform the same amounts to their general and subjective standards. Even non-accountants in your Organization will understand an hourly moving average that highlight Sales, Cash Flow & Operating Earnings; no different from what you pay all your employees as a composite average.**

The many hours devoted to Budgeting, Reporting, and Analyzing Results will be greatly reduced from the mere fact that all of Management will be observing the same variations to Unit Sales detail. All Financial Analysis will be driven by Long Range Planning (for what ever duration) that is never confronted, but is always apparent by an Operating Budget each current year.

The best part, the data is already alive while only a matter of Accounting receiving daily reports from Human Resources and Sales. No current reporting will change with Management getting a weekly spreadsheet of Performance Cash that is penny sensitive and reconcilable to Operating Earnings at every turn.

Sincerely,

Managing Partner XYZ Partners

Martin D'Amico

About the Author

As Corporate Controller, he vividly remembers Amerace Corporation and its 13% interest costs that had to be satisfied by a steady stream of cash flow. Ignoring weekly customer demand as reflected by each week's cash flow in favor of reporting monthly Operating Earnings with all its accounting adjustments is depriving a business man of the most consistent and unmitigated truth of a business.

It is a rare General Manager or CEO that understands the relationship of cash flow to operating earnings; it is not easy, nor necessary. Concentrate on cash as reflected by sales and a check book, when flow remains positive and a Manager knows the ultimate sources; Operating Earnings will fall into place.

As a Manager you are well aware of what depletes cash, it is time to learn more about its most significant and volatile sources 1) how many units a business sells 2) the price of each different major product. D'Amico first brought unit knowledge to the business world as a Manufacturing Controller back in the 80's when he converted unit production reporting for a multi product plant to hours. How many **different units** were produced or sold was never an Amerace problem again.

In January 1999 Management Accounting Magazine published and validated D'Amico's unit selling price relationship to **"hours paid"** for its entire Accounting Membership. In November of the same year Public Accounting Magazine addressed their Professional readers with similar legitimization. In both articles selling prices were linked to hours and cash flow for the benefit of the Entrepreneur. Although cash flow verse earnings remained reconcilable, Accountants as a group of financial advisors, felt no need to pass hour **"know how"** on to internal management or their clients.

The deliberation by Accounting Organizations over **the hour/cash flow concept** allowed him to perfect an idea instigated in the Public Accounting review of his first book ---a series of different "hands–on and real–time" excel sheets itemized in the Table of Contents. An accountant's endorsement as to how it is done is no longer necessary, only a General Manager's desire to have it done. There is an Excel Worksheet in the Appendix that includes cash flow in the traditional sense. All Excel Sheets are applicable to Operating Earnings as well.

Even when sales dollars are equal to or surpass those planned, variations to earnings are burdensome to account for, not timely, and in a great many cases related to Accounting journal entries. **Weekly Performance Cash** cuts out the journal entries so that Management can influence the future with simple reasoning. If the principal product, 10,000 hamburgers or widgets have been selling for $4.00 over a 7 month period it is not difficult to predict cash flow on those assumptions for the remainder of the year.

Realizing that many readers would not have a working knowledge to use Excel charts & a diskette, the book was first written with charts to aid readers in understanding Earned Hours© and the 100% process. The chart files in the text were added next to ease the education process using the hands on capability of Excel. To aid users with real time reports that can be used in business next week, the Appendix Files (XL) were added last. Any author who writes on technical subjects such as Finance or Accounting without utilizing the powerful resource of links and the variability of files or programs is not serving the reader adequately. A picture is worth a 1000 words & it is the reason for the abbreviated text.

Most charts and XL files were text copied with the cell references in tact; in case there are any specific questions they can be answered via Email---www.performancecash.com

I will not leave you in a lurch when using Earned Hours immediately. If you insert one week of complete data for your business on any one of Appendix Files and send via e mail I will return a fully programmed report that you can use for the balance of your current year. All you need to tell me--- what week in your calendar or fiscal year it represents and your Year to Date net sales dollars consistent with that week.

This arrangement has allowed me to price the book for the consumer audience. In addition Performance Cash is not meant to be read in its entirety but any one person, only relative to each reader's specific need.

Special Private Label Order Form for Performance Cash

The strength of **Weekly Performance Cash lies within its accumulation from & for each operating facility.** Its power is in the **author's permission to reproduce the entire contents** of the **Diskette. Such permission to use, or copy & distribute to each of your facilities or clients will get the process started with a mere guess for subsequent & accurate cash flow forecasting within a short time. The use of the book's four benchmarks will not change but augment any reporting currently in place.**

Your end result, **you or your client** will receive **One Weekly Excel Spreadsheet** from each operating entity. Eventually the weekly report will **first relate** the cause of any variance as to Sales, Cash Flow & Earnings compared to the latest expectations. After which you as Accountant or your clients **will be in a position to accumulate the results <u>for an entire Corporation</u>.**

With a single copy purchase you are entitled to downlosd one copy of the Appendix Diskette from (http://www.performancecash.com/) at a price commensurate with Earned Hours© of the text. The Diskette with its Excel folder contains 20 different self Contained Programs & applications referenced in the Table of Contents of this book & described in the Appendix.

If you want to arrange for multiple copies of the Diskette *you may reproduce your own copies in larger quantities as outlined below. The specific prices for the special order can be obtained at www.performancecash.com.*

No instructions will be necessary when distributing the Diskette within your Accounting organization— **only this comment.**
"Please have your first Performance Cash Report & Your Weekly Break Even on my desk in two weeks; there are no incorrect presentations of the Truth & no better Control of Your Business than a Break Even Based primarily upon units sold & secondarily on costs.

Up to 100 copies of the Appendix Diskette
Up to 500 copies of the Appendix Diskette
Unlimited copies of the Appendix Diskette

**Any Diskette Arrangement entitles you to receive the author's other specifically designed Corporate Year to Date Excel Worksheets (XL 100 Series). They were intentionally excluded from the Appendix Diskette to keep the price within the means of a business just starting out. The author plans to add programs to the XL 100 Series as users customize & enhance the Earned Hour Concept.

***In addition Conference Call Seminars with Visuals on your screen (it is that simple) are available.

Enclosed is my check for $_____ to make multiple copies of the Diskette. I understand & I will abide by the reproduction quantities ordered above.
Name_____ Title_____ Firm_____

Address_____

E-mail_____ Phone_____ Signature_____

Any purchase may also be charged via my web-site *performancecash.com.*